CROP PRODUCTION PRACTICES

CROP PRODUCTION PRACTICES

By
Oleg Ishutin

2016

SBS Publishers & Distributors Pvt. Ltd.
New Delhi

ISBN 13 : 9789380090689

First Published in 2016

Published by:

SBS PUBLISHERS & DISTRIBUTORS PVT. LTD.

2/9, Ground Floor, Ansari Road, Darya Ganj,

New Delhi - 110002,

INDIA

Tel: 0091.11.23289119 / 41563911

Email: mail@sbspublishers.com

www.sbspublishers.com

Preface

Crop production is a complex business, requiring many skills and covering a variety of operations throughout the year. Crop production systems that require chemical fertilizers, pesticides, machinery for tillage, and irrigation water are expensive. They have started to undermine the water security of future generations, contributing to soil and water pollution particularly when synthetic pesticides are not used properly. It is true that agriculture as practiced 100 years ago without modern inputs had lower productivity than present systems of production. However, many pre-modern practices, such as the use of organic manures to enhance soil fertility and of herbal extracts to protect crops, can be made more efficient by the scientific knowledge that has been gained over the past century, making crop production more sustainable while still achieving high productivity. This is becoming more evident from the published literature on practices such as the use of organic manures and bio-pesticides and experience with conservation tillage. A crop is any cultivated plant, fungus, or alga that is harvested for food, clothing, livestock fodder, biofuel, medicine, or other uses. In contrast, animals that are raised by humans are called livestock, except those that are kept as pets. Microbes, such as bacteria or viruses, are referred to as cultures. Microbes are not typically grown for food, but are rather used to alter food. Crop production is a core business for the majority of agricultural holdings world-wide and is the sole source of income for many growers. This short-course module, suitable for graduates and non-graduates working in the Crops Industry, who are planning to become growers or technical advisors, aims to provide an understanding of the scientific principles underlying crop production and their application in commercial crop management.

Editor

Contents

Chapter 1

EFFECTS OF INCREASED NITROGEN DEPOSITION AND ROTATION LENGTH ON LONG-TERM PRODUCTIVITY OF CUNNINGHAMIA LANCEOLATA PLANTATION IN SOUTHERN CHINA

[1]Meifang Zhao, [1]Wenhua Xiang, [1]Dalun Tian, [1]Xiangwen Deng, [1]Zhihong Huang, [2]Xiaolu Zhou,[1,2] Changhui Peng

[1]Faculty of Life Science and Technology, Central South University of Forestry and Technology, Changsha, Hunan, People's Republic of China

[2]Institute of Environment Sciences, Department of Biological Sciences, University of Quebec at Montreal, Montreal, Quebec, Canada

ABSTRACT

Cunninghamia lanceolata (Lamb.) Hook. has been widely planted in subtropical China to meet increasing timber demands, leading to short-rotation practices that deplete soil nutrients. However, increased nitrogen (N) deposition offsets soil N depletion. While long-term experimental data investigating the coupled effects related to short rotation practices and increasing N deposition are scarce, applying model simulations may yield insights. In this study, the CenW3.1 model was validated and parameterized using data from pure *C. lanceolata* plantations. The model was then used to simulate various changes in long-term productivity. Results indicated that responses of productivity of *C. lanceolata* plantation to increased N deposition were more related to stand age than N addition, depending on the proportion and age of growing forests. Our results have also shown a rapid peak in growth and N dynamics. The peak is reached sooner and is higher under higher level of N deposition. Short rotation lengths had a greater effect on productivity and N dynamics than high N deposition levels. Productivity and N dynamics decreased as the rotation length decreased. Total productivity levels suggest that a 30-year rotation length maximizes productivity at the 4.9 kg N ha^{-1} $year^{-1}$deposition level. For a specific rotation length, higher N deposition levels resulted in greater overall ecosystem C and N storage, but this positive correlation tendency gradually slowed down with increasing N deposition levels. More pronounced differences in N deposition levels occurred as rotation length decreased. To sustain *C. lanceolata* plantation productivity without offsite detrimental N effects, the appropriate rotation length is about 20–30 years for N deposition levels below 50 kg N ha^{-1} $year^{-1}$ and about 15–20 years for N deposition levels above 50 kg N ha^{-1} $year^{-1}$. These results highlight the importance of assessing N effects on carbon management and the long-term productivity of forest ecosystems.

INTRODUCTION

Nitrogen (N) is the element that has the greatest limiting effect on plantation productivity [1]. However, in many areas, forest ecosystems have experienced increased atmospheric N deposition in recent years [2]-[5]. For example, in most forests in subtropical China, the current rate of N deposition ranges from 18 to 73 kg N ha^{-1} $year^{-1}$, and this is expected to increase in coming decades [6]-[9]. The increase in N that is deposited in forests will increase soil N content and, consequently, stimulate increased productivity over the short term [10]. Although N deposition could meet the N requirements of these forests, excess N will result in a nutrimental burden and offsite environmental effects [5], [11], [12]. Balancing the N supply and the demand for tree growth in these forests is critical for optimizing productivity, sustainability, and environmental protection [13], [14].

N cycling in forest ecosystems is a complex process because of the numerous soil-plant interactions. Nutrient pools and fluxes, as well as the nutrient requirements of the forest trees, vary as the stands develop [15]. To determine the effects of management practices on nutrient cycling and the responses in forest productivity that occur following changes in N input, long-term observational data sets are required. Previous studies have manipulated N levels to examine the short-term responses in tree growth and assess N requirements. However, these studies did not consider variations in the long term productivity and the N requirements of forest stands as they develop. Mechanistic models can accurately predict long-term changes in productivity and are widely used in N fertilizer applications, resulting in increased N-use efficiency and reduced pollution [15], [16]. Some mechanistic models have been used to analyze the effects of rotation length on tree and soil carbon (C) stocks and the quality of wood in different European forests [17], the effects of harvesting intensity and rotation length on long-term soil N and C dynamics in boreal forests in central Canada [18], the effects of N deposition, climate change, and rotation length on forest carbon sequestration and the harvesting of Scots pine forests in southern Finland [19], and to compare the ecological impact of natural disturbances and harvesting [20].

Cunninghamia lanceolata (Lamb.) Hook. is the third most commonly planted tree species in plantations in the world [21], and its total plantation area is now 9.21 million hectares in China[22]. Due to its high value in terms of timber quality and versatility, *C. lanceolata* has mainly been planted for timber production, but its bark and roots are usually harvested for local construction purposes and fuel [23], [24]. Currently, most *C. lanceolata* plantations are monocultures and successive rotation planting is commonly practiced [24], [25]. Increased timber demand and improvements in processing techniques have resulted in shorter harvesting cycles of *C. lanceolata* plantations (originally 30 years to now $\leq$20 years). Soil N depletion due to timber removal has been recognized as the major factor responsible for the decline in yields of *C. lanceolata* plantations [26]. Therefore, sustaining the productivity of *C. lanceolata*plantations over successive rotations is a concern of numerous researchers [27]–[29] and has been identified as a priority for plantation management in China [4]. Regarding *C. lanceolata* plantations, various models have been used to simulate age-related C dynamics [30], [31], yield decline due to management activities and degradation of soil fertility [28], and C sequestration at N deposition levels <1–50 kg N ha^{-1} $year^{-1}$ at various rotation cycles [32]. However, the simultaneous effects of increased N deposition and rotation lengths with successive planting on long-term production and management in *C. lanceolata* plantations have not been studied so far.

This study uses the CenW3.1 model [33], [34] to simulate long-term production and N requirement dynamics in response to N deposition and different rotation lengths and establish sound management practices for the sustainable productivity of *C. lanceolata* plantations. The major parameters of production and N dynamics include stem wood biomass, net primary production (NPP), annual N requirement (ANR) for tree growth, soil available N for growth, and soil organic C (SOC). Specifically, the purposes of this study are to (1) determine how increased N deposition affects forest productivity as stand development; (2) examine the interactive effects of N deposition and rotation length on long-term forest productivity; (3) investigate whether practicing increased atmosphere N with shorter rotation that can avert N saturation damage in the study region.

Materials and Methods

Site Description

The *C. lanceolata* plantations that were evaluated in this study are located at the experimental area of the Huitong National Forest Ecosystem Research Station (NFERS) (lat 26°50′N, long 109°45′E), Hunan Province, southern China [31]. The region is characterized by a humid mid-subtropical monsoon climate. The annual mean temperature is 16.5°C, which ranges from an average of 4.3°C during the coolest month (January) to 29.4°C during the warmest month (July). The annual rainfall is approximately 1270–1650 mm, mainly occurring between April and August. The soil is a subgroup of clay loam red soil that is formed from shale and slate parent materials (Table 1).

Table 1. Soil characteristics of the C. lanceolata plantations at the Huitong National Forest Ecosystem Research Station, Hunan Province, China.

Characteristic	Value	Source
Soil pH	4.32–4.86	[60]
Bulk density (g cm^{-3})	1.09–1.7	[61]
Porosity (% of the total volume)	52.1–62.5	[62]
Permeable velocity (cc min^{-1})	2.83–10.1	[62]
Maximum moisture holding capacity (% of the total volume)	36.1–40.4	[62]
Moisture content (%) for most of the year	>30	[63–64]
Total N (g kg^{-1})	1.876–2.478	[65]
NH_4^+-N (mg kg^{-1})	6.67–9.1	[66]
NO_3-N (mg kg^{-1})	11.2–24.3	[67]
Total P (g kg^{-1})	0.92–1.42	[68]
Total K (g kg^{-1})	4.27–5.13	[69]
TOC (g kg^{-1})	21.1–33.4	[70]
Soil C:N ratio	13–16	[64]

doi:10.1371/journal.pone.0055376.t001

doi:10.1371/journal.pone.0055376.t001

Eight *C. lanceolata* plantation watersheds were established within the research station in 1984. Automatic weather stations were installed within the plantations to collect climatic data. The first-rotation *C. lanceolata* plantations were established on a clear-cut of natural forest in 1966. After the slashed understory plants and harvesting residues had been burned, the soil was prepared by digging holes and 1-year-old *C. lanceolata* seedlings were planted. No weed control, fertilization or other treatment was employed. The establishment, survival, and growth of the plantation have been described previously [31]. This study was carried out at two watersheds (II and III), situated within NFERS. The second successive rotation plantation was established in watershed III in February 1987 and in watershed II in March 1995 with an initial planting density of 3318 trees ha^{-1} using identical planting methods as used in the first rotation. The ecological factors and characteristics of the *C. lanceolata* plantations related to this study were previously described [31]. The biomass and stand characters, including height (H, m), diameter at breast height (DBH, cm), and stand density (trees ha^{-1}), were measured at different growing-stages beginning at age 7. Total volume (under bark) and dry matter biomass of the wood, bark, branches, and needles were calculated. Biomass, litter, forest floor, and understory samples were collected at an interval of five years for nutrient concentration and growth analysis. Linear regression was used to test the relationships between soil bulk density and soil C content.

CenW Model

The CenW3.1 model was developed as a comprehensive monoculture forest model that runs in daily time-step with the C gain that is calculated from light absorption and can be modified by taking into account temperature, soil water status, and foliage N concentration [34]. CenW allows the explicit modeling of both C and N pools as well as their fluxes, thereby integrating tree growth parameters, environmental factors, and management treatments. The projection of stand growth and N dynamics are based on the rates of key ecological processes regulating the availability of and competition for light and nutrient resources. Net primary productivity (NPP, t ha^{-1} $year^{-1}$, represents the remains of stand-level photosynthesis after subtracting losses of stand-level respiration and litterfall) can be reduced through a number of processes, such as increasing respiration, increasing senescence and mortality losses, immobilisation of nutrients and unfavourable shifts in biomass allocation. Fixed photosynthate is used for plant growth, with allocation to different plant organs determined by plant nutrient status, tree height and species-specific allocation factors. Stem wood biomass (t DM ha^{-1}) includes heartwood and sapwood. Tree death is estimated as a simple daily fractional mortality rate. It is assumed that the ratio of above- to below-ground allocation increases with foliar N concentration. Foliar N concentrations are essentially determined through the relative rates of C and N uptake. It is also assumed that 25% of foliar N is relocated to other plant part before litter fall. Water use is calculated using the Penman-Monteith equation, with canopy resistance given by the inverse of stomatal conductance, which, in turn, is linked to calculated photosynthetic carbon gain. Water is lost through transpiration and soil evaporation, and gained by rainfall or irrigation which together determines soil water status for the following day. Stand density was fully independently modeled as a result of self-thinning. The nutrient cycle is considered through litter production by the shedding of plant parts, such as roots, bark, branches and foliage. Litter is assumed to be produced as a constant fraction of live biomass pools. In addition, foliage is shed during drought or when canopies become too dense. Litter is then added to the organic matter pools from where C is eventually lost and N becomes available again as inorganic mineral N. N can come from external addition (atmospheric deposition and fertilizer addition) or mineralization during the decomposition of soil organic matter. Plant available N depends on is current soil available N status plus additions from atmospheric deposition. Soil decomposition rate is determined by temperature, soil water status and soil organic matter quality. Input litter tends to have a wide C:N ratio so that some C needs to be lost through decomposition before critical C:N ratio is reached in organic matter and excess N can be mineralized. The model also offers the potential to manipulate forests in a variety of experimental ways and predict outputs. Detailed descriptions of model features, structures, mathematical representations, sensitivity analyses, and building strategies have been described [33], [34].

Model Calibration and Validation

The calibration and validation of CenW3.1 in *C. lanceolata* plantations in this study followed two steps. The first step for model calibration was to select the data from watershed II to calibrate CenW3.1 model. The purpose of this step was to determine the model parameters to ensure simulation accuracy as much as possible. The second step for model validation was to evaluate the model against the independent watershed III to test the model accuracy. The data used in the model simulation included environmental factors, tree growth parameters, tree-N response data, forest floor decomposition rates, and N inputs. The primary parameters and values for productivity, above-ground biomass, decomposition, and soil C were previously described in detail [31] (see Appendix S1).

For this study, the CenW3.1 simulations were run for a period of 120 years using climate records that were obtained during the period from 1987 to 2010. The historic records of daily maximum and minimum values of air temperature, daily sum of radiation (MJ m^{-2} day^{-1}), and daily precipitation (mm) over the period 1987–2010 at the study site were used for climate inputs. For future periods without observed climate data, we randomly assign each year's climate condition with detrended climate data [35].

Tree growth parameters, tree-N response data, forest floor decomposition rates, and N inputs were obtained from the literature or field measurements (see Appendix S1). The rates of these processes were calculated from a combination of historical bioassay data (biomass accumulation in component pools, stand density) and measures decomposition rates and photosynthetic saturation curves by relating 'biologically active' biomass components (foliage and small roots) to calculate nutrient uptake, the capture of light energy, and net primary production. Then, the model generated a suite of growth properties for each tree components to be represented based on the historical data (i.e. growth and yield tables or long-term permanent plots), which was used to calibrate tree growth. These growth properties were subsequently used to simulate plant growth as a function of resource availability and management practices. Calibration data were assembled that describe the accumulation of tree biomass (above and below-ground components) as stands developed. Aging was taking into account by using calibration data pairs for different stand ages (i.e. age vs. tree size, age vs. stem density, age vs. biomass, etc.). Tree biomass and stand self-thinning rate data were estimated from the height, DBH and stand density as output of traditional empirical growth and yield models in conjunction with species-specific component biomass allometric equations. To calibrate the nutritional aspects of the model, data describing the concentration of nutrients in the various biomass components were required. The data required in the model on the degree of shading produced by different quantities of foliage and the response of foliage to different light levels were derived from literature values, field measurements, or simulation models. Lastly, decomposition rates of various litter types and soil organic matter were required for the model to simulate nutrient cycling. Fertilizer applications, irrigation management techniques, weather conditions (such as sunny, cloudy, and etc.), incidences of insect problems or disease, plowing, and fire were excluded in the simulations.

Moreover, the annual N requirement (ANR) for tree growth was defined as the year-to-year change in amounts of N required for biomass accumulated during the year (expressed in kg N ha^{-1} year^{-1}). This was estimated using the following equation:

$$ANR_t = \sum_{i=1}^{n} (B_{i(t)} \times C_{i(t)} - B_{i(t-1)} \times C_{i(t-1)}) \quad \textbf{(1)}$$

where ANR_t is the annual N requirement at stand age of year t; $B_{i(t)}$ and $B_{i(t-1)}$ represent the accumulated biomass of tree compartment i at stand age of year t and $t-1$, respectively; $C_{i(t)}$ and $C_{i(t-1)}$ represent the N concentration in tree compartment i at stand age of year t and $t-1$; and n is the number of compartments (including sapwood, heartwood, bark, branches, leaves, fine root, coarse root, flower, and fruits) within the stand. B_i was quantified as the amount of biomass in each compartment i, plus the biomass returned to the soil as litter, predicted by CenW3.1 for each stand during each rotation cycle.

Linear regressions of the observed data obtained from the study site were compared with the simulated average stand height (H), average diameter at breast height (DBH), stand density (stems per hectare), and stem woody dry matter biomass that were calculated using CenW3.1. These comparisons were used to validate the CenW3.1 model for long-term predictions (Fig. 1). For a better quantification of the error in fitting low flows, the efficiency proposed by Nash and Sutcliffe (NSE) [36], defined as one minus the sum of the absolute squared differences between the predicted and observed values normalized by the variance of the observed values during the period under investigation, was used to tested modeling performances.

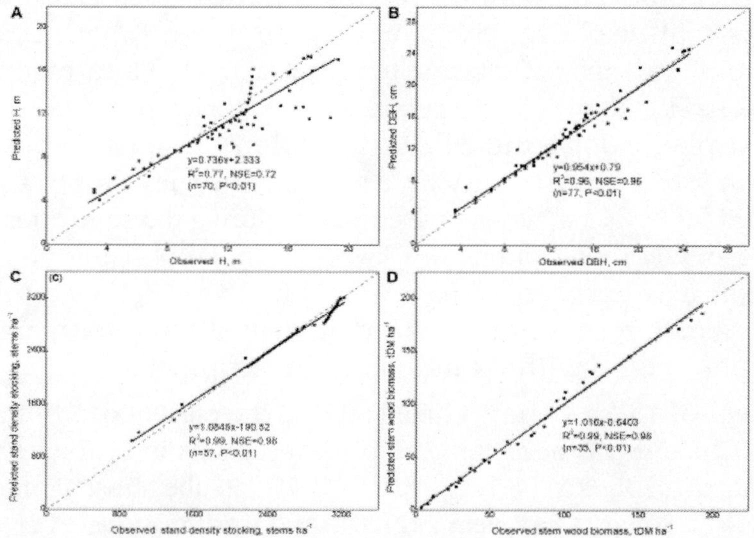

Figure 1. The model was validated by fitting the predicted and observed values. (A) Stand height (H, m). (B) Diameter at breast height (DBH, cm). (C) Stand stocking (stems per ha). (D) Stem woody biomass in tone dry matter per hector (t DM ha^{-1}). Gray dashed lines represent a 1:1 relationship; solid lines are linear regressions. NSE for "Nash-Sutcliffe efficiency".

doi:10.1371/journal.pone.0055376.g001

The H value (Fig. 1A) that was calculated by the model fitted well with the observations across different climatic conditions ($R^2 = 0.77$, NSE = 0.72). The predicted and observed data were closely matched for DBH (Fig. 1B; $R^2 = 0.96$, NSE = 0.96), stand density (Fig. 1C; $R^2 = 0.99$, NSE = 0.98), and stem woody dry matter mass (Fig. 1D; $R^2 = 0.99$, NSE = 0.98) in C. *lanceolata*plantations during the period 1990–2010. The validation results, therefore, confirm the reliability of the applied model for further investigations into the long-term effects of N deposition and rotation lengths on the productivity and N dynamics of C. *lanceolata* plantations.

Model Simulation Runs

Five dynamic variables including stem wood biomass, NPP, ANR, soil available N for growth (kg N ha^{-1}), and soil organic C (SOC, t C ha^{-1}), were consider in this study. However, these variables are sometimes not sufficient to account for the full magnitude of changes. This study combines that response with simple assumptions about external N depositions and current rotation length practices. The atmospheric N depositions (kg N ha^{-1} $year^{-1}$) refers to external input rates as wet and dry atmospheric deposition. For effects of increased N deposition on long-term forest productivity, five N deposition levels e.g. 4.9, 18, 30, 50, 70 and 90 kg N $ha^{-1}year^{-1}$ were applied during the simulations. Based on the reported N depositions in China, the initial N deposition in 1987 is approximately estimated to be 4.9 kg N ha^{-1} $year^{-1}$ (N4.9) at the study site [37], which represents the average deposition rate in areas with human population without industrial development.

A level of 18 kg N ha^{-1} $year^{-1}$ (N18) is the current deposition rate in southern China [8]. Whereas in the polluted areas in Hunan Province, levels around 30 kg N ha^{-1} $year^{-1}$ (N30) is the most common, but reaches up to 50 kg N ha^{-1} $year^{-1}$ (N50) [8]. Higher levels over 70 kg N ha^{-1} $year^{-1}$ (N70) in the large industrialized urban areas have been reported. It could further increases up to 90 kg N ha^{-1} $year^{-1}$ (N90) for near future [4]. In the model, it was taken to be a constant rate throughout the life of the stand evenly distributed into every day. Assuming no major climatic changes, the atmospheric N depositions alone is relatively constant for a given site over time. We will also assume that it is constant within a given rotation and over successive rotations. Harvesting and thinning affect C and N dynamics in different ways, depending on the harvesting intensity and the rotation length. In this study, C and N dynamics were simulated and compared using rotation lengths of 120 (RL120), 40 (RL40), 30 (RL30), 20 (RL20) and 15 (RL15) years over a period of 120 years. All rotations started with the planting of 3318 seedlings per hectare. At the end of each rotation period, the stands were clear-felled, and bole only harvesting was adopted, the stem wood at harvest was considered as removed from the site. When a bole harvesting only event occurs, 94% the wood and bark N and volume were removed from the site. The remaining 6% is transferred to the forest floor. The rate of removal of foliage and branch N and volume from the site remains at zero, since the entire amount is transferred to the forest floor as slash. The long-term growth was seldom affected by residue treatment and field observations generally confirm this [38].

The model was used to simulate growth (physiological) constrains on long-term productivity in the study site over a period of 120 years without the effects of rotation (RL120) under initial level of N deposition (N4.9). The growth responses to increased N deposition were re-expressed as growth-response (ecophysiological) constraints on long-term productivity over a period of 120 years without rotation (RL120) under each increased deposition level (N18, N30, N50, N70 and N90).

The model was then used to investigate the effects of rotation lengths (RL40, RL30, RL20, and RL15) on long-term forest productivity for initial (N4.9) or increased (N18, N30, N50, N70 and N90) N deposition. We examine the changes in long-term forest in response to different rotation length and increased N deposition, and figure out an appropriate rotation length for maximizing total stem biomass and average NPP. This approach did not address the issues of biological and genetic diversity conservation, or wider socioeconomic aspects of sustainability, such as energy costs associated with fertilizer inputs, harvest operations, transport and processing of wood products.

Results

Effects of Increased N Deposition

The simulated productivity and N dynamics (Panel A, B, C in Fig. 2) of a *C. lanceolata*plantation, in terms of stand growth under the baseline (N4.9 scenario; Fig. 2a0, b0, c0) and increased N deposition impacts (N18, N30, N50, N70 and N90 scenarios; Fig. 2a1–5, b1–5, c1–5) for a period of 120 years, are presented in Fig. 2.

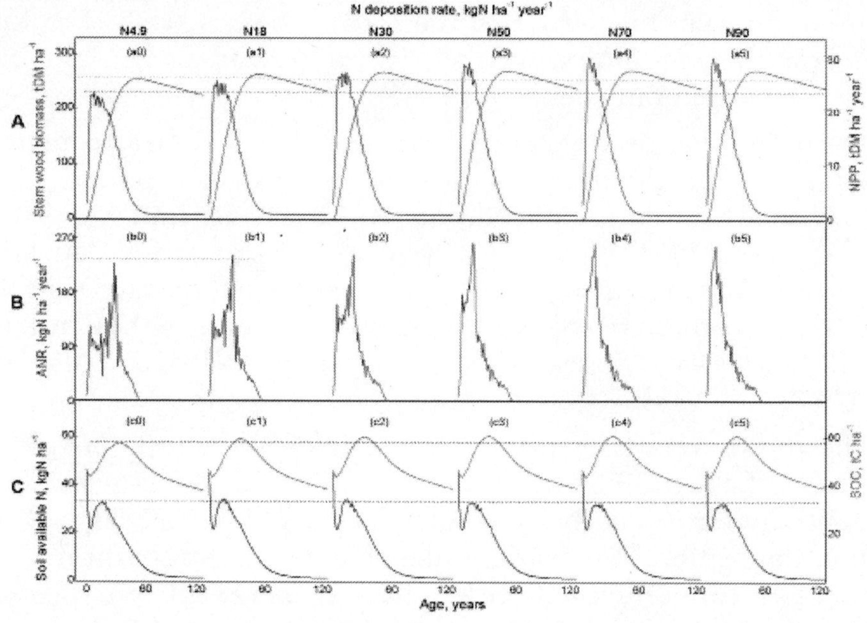

Figure 2. Simulated increased N deposition impacts on forest productivity as stand development. (Panel A) The magenta solid line stands for the stem wood biomass production in tone dry matter per hector (t DM ha^{-1}), and green solid line refers to the net primary production in tone dry matter per hector (NPP, t DM ha^{-1} $year^{-1}$). (Panel B) Annual increment of ANR (red solid line, ANR, kg N ha^{-1} $year^{-1}$). (Panel C) The blue solid line represents current status of soil available N for growth in kg N per hector (kg N ha^{-1}), olive solid line refers to soil organic C (SOC, t C ha^{-1}) of *C. lanceolata* plantation in southern China from seedling to 120 years of age. In panels A, B, and C, a0, b0 and c0 represent growth trends from the basic, under initial level of N deposition (N4.9) and

without rotation practice (RL120), simulation that made up the baseline data set; and a1–5, b1–5, as well as c1–5 represent predicted growth response trends generated under increased N deposition. The dashed lines represent growth constraint equilibrium state.

doi:10.1371/journal.pone.0055376.g002

The baseline simulation (Fig. 2a0, b0, c0) generated a realistic sigmoid function of age in the general pattern of stand growth in which the total dry matter of stem wood biomass continuously increased with stand development, peaked at 255.84 t DM ha^{-1} at a stand age of 53 years, followed by a gentle decline as the stands grew, and yielded at 226.35 t DM ha^{-1} at the end of 120-year period (Fig. 2a0). The increase in wood production with stand age was corresponding to the "growth expansion first and decline later" pattern of NPP (Fig. 2a0). The decline trends as stand grows after 53 years can be explained from ANR (<0 after 54 years) (Fig. 2b0). NPP initially exhibited a steep increase during the first 10 years, approximated a maximum rate (here 22.08 t DM ha^{-1} $year^{-1}$) at 13 years, and then gradually declined with growth. Thereafter, NPP decreased to nearly 0 (here 0.3 t DM ha^{-1} $year^{-1}$) at 60 years. Decreased nutrient availability and enhanced stomatal limitation were the major causes for NPP decline with stand age. Possible shifts in allocation to belowground components also may contribute to the apparent decline in productivity and biomass accumulation rate of stem wood. The average NPP over the 120-year period was 5.78 t ha^{-1} $year^{-1}$ (Appendix S2).

Annual N requirement (ANR) demonstrated a drastically different pattern compared with production as the stand developed, formed an age-related "concave up beginning convex later" trends (Fig. 2b0). In terms of increases in foliage and fine root pools, ANR increased sharply from 9 kg N ha^{-1} $year^{-1}$ at 2 years to a maximum value of 109.6 kg N ha^{-1} $year^{-1}$ at 16 years. Then caused by self-thinning, ANR generally decreased between 16 and 23 years of age, corresponded to NPP fluctuates at this period. During the period from 23 to 29 years, ANR increased from 70.7 kg N ha^{-1} $year^{-1}$ to a maximum of 227.9 kg N ha^{-1} $year^{-1}$. After 29 years, ANR decreased and fell to <0 after 54 years, which indicates any increase in stand level respiration, hydraulic resistance, and perhaps plant tissue maturation as age grew. The average ANR over the 120-year period was 32.4 kg N ha^{-1} $year^{-1}$(Appendix S2).

Soil available N (Fig. 2c0) decreased from 55.9 kg N ha^{-1} at the time when the seedlings were planted to 21.5 kg N ha^{-1} at a stand age of 5 years (Fig. 2c0). As the stand developed, soil available N increased to 32.6 kg N ha^{-1} during the period from 5 to 15 years. Thereafter, soil available N decreased to 1.5 kg N ha^{-1} at 120 years of age. Similar to soil available N, SOC (Fig. 2c0) varied over the first few years after harvest and then declined by about 10% for the next 5 years (Fig. 2c0). SOC dropped from 48.3 t C ha^{-1} at the time when the seedlings were planted to 43.6 t C ha^{-1} at a stand age of 5 years, and then returned to the baseline level by a stand age of 15 years. Then, SOC increased reaching a peak value of 57.4 t C ha^{-1} at an age of 35 years. Thereafter, SOC decreased to 37.8 t C ha^{-1} at 120 years.

Fig. 2 a1–5, b1–5, c1–5 show the variation of stem wood biomass (t DM ha^{-1}), NPP (t DM $ha^{-1}year^{-1}$), ANR (kg N ha^{-1} $year^{-1}$), soil available N (kg N ha^{-1}) and SOC (t C ha^{-1}) with stand ages, predicted by the model under increased N deposition rate (without the influence of harvesting regimes) over the 120 year study period. The results showed that productivity of *C.lanceolata* plantation in response to increased N deposition was more related to age than N addition depending on the proportion and age of stand development. Increased N deposition has shown a rapid peak in growth and N dynamics. Increasing N depositions increased average NPP to 5.95, 6.032, 6.05, 6.06, and 6.06 t DM ha^{-1} $year^{-1}$ under the N deposition rates of 18, 30, 50, 70 to 90 kg N ha^{-1} $year^{-1}$, respectively (Appendix S2). Increased available N, decreased stomatal limitation and early arrival of a large leaf area were the major causes for the growth peaks decline with stand age.

Effects of Rotation Lengths Under the N4.9 Scenario

Under the null level of N deposition (N4.9; gray solid lines in Fig. 3, solid squares in Fig. 4, and pink solid line with hollow squares in Fig. 5), the effects of rotation lengths on the long-term productivity of *C. lanceolata* plantation are shown in Figs. 3,4,5 and Appendix S2.

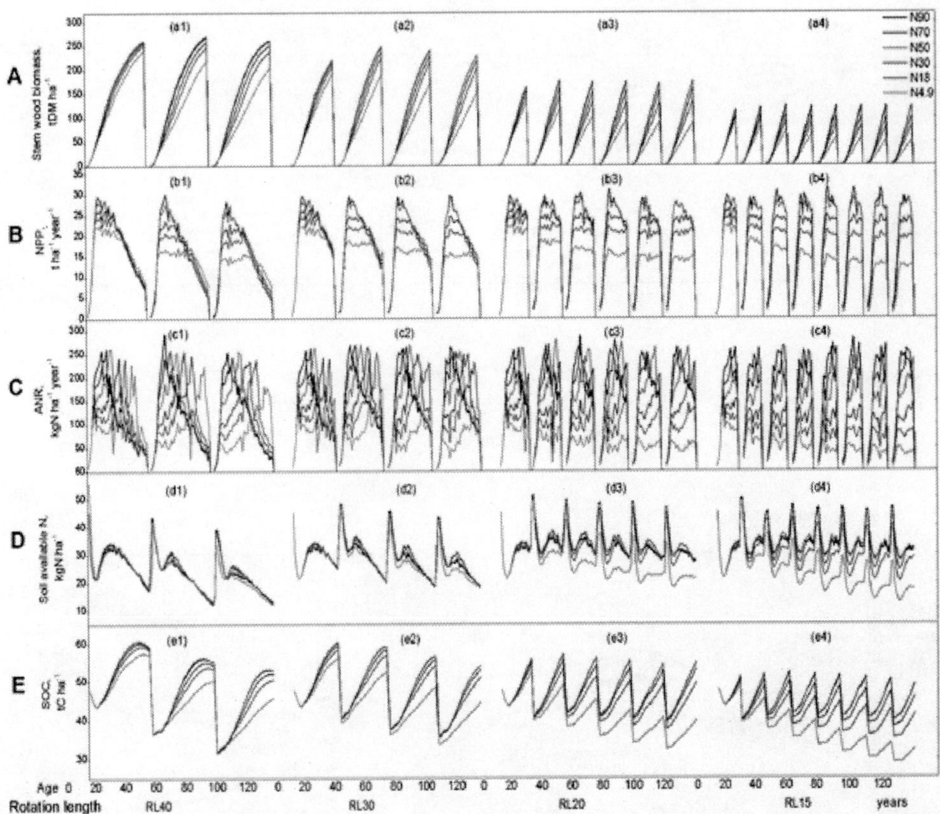

Figure 3. Simulated interactive effects of increased N deposition and short-rotation management as stand development. (A) The stem wood biomass production in tone dry matter per hector (t DM ha^{-1}). (B) The net primary production in tone dry matter per hector (NPP, t ha^{-1} $year^{-1}$). (C) Annual increment of ANR (kg N ha^{-1} $year^{-1}$). (D) The current status of soil available N for growth in kg N per hector (soil available N, kg N ha^{-1}). (E) Soil organic C (SOC, t C ha^{-1}). Over the 120-year period coupling chronic levels of atmosphere N depositions of N4.9, N18, N30, N50, N70, and N90 with different rotation intervals (RL40, RL30, RL20, and RL15). RL40, RL30, RL20, and RL15 represent rotation cycles with intervals of 40, 30, 20, and15 years, respectively. N4.9 (black solid line) = 4.9 kg N ha^{-1} $year^{-1}$, N18 (gray solid line) = 18 kg N ha^{-1} $year^{-1}$, N30 (light gray solid line) = 30 kg N ha^{-1} $year^{-1}$, N50 (black dash) = 50 kg N ha^{-1} $year^{-1}$, N70 (gray dash) = 70 kg N ha^{-1} $year^{-1}$, and N90 (light gray dash) = 90 kg N ha^{-1} $year^{-1}$.

doi:10.1371/journal.pone.0055376.g003

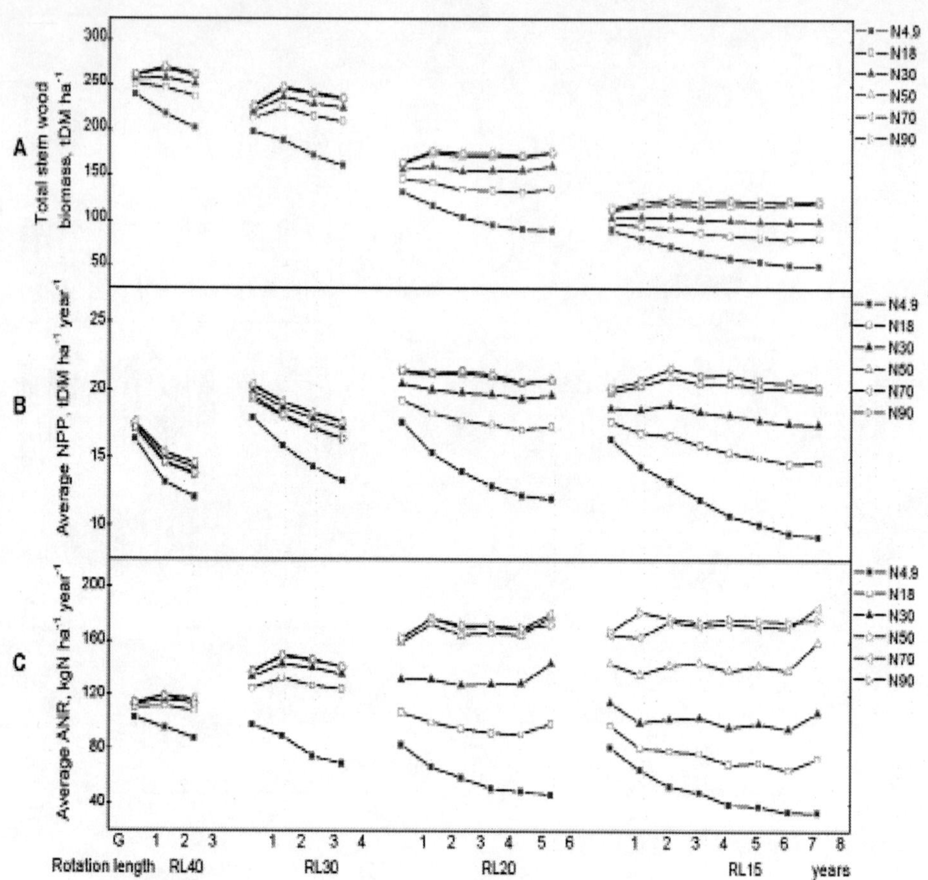

Figure 4. The summary of simulated interactive effects of increased N deposition and short-rotation management over successive rotations. It displays the accumulative statues, (A) Stem wood biomass at the end-of-rotation, and rotation-average values for flux or rates, (B) NPP and (C) ANR. RL40, RL30, RL20, and RL15 represent rotation cycles with intervals of 40, 30, 20, and 15 years, respectively. The effects of atmosphere N deposition levels: N4.9 (solid squares) = 4.9, N18 (hollow squares) = 18, N30 (solid upward triangles) = 30, N50 (hollow upward triangles) = 50, N70 (hollow leftward triangles) = 70, and N90 (hollow rightward triangles) = 90 kg N $ha^{-1}year^{-1}$.

doi:10.1371/journal.pone.0055376.g004

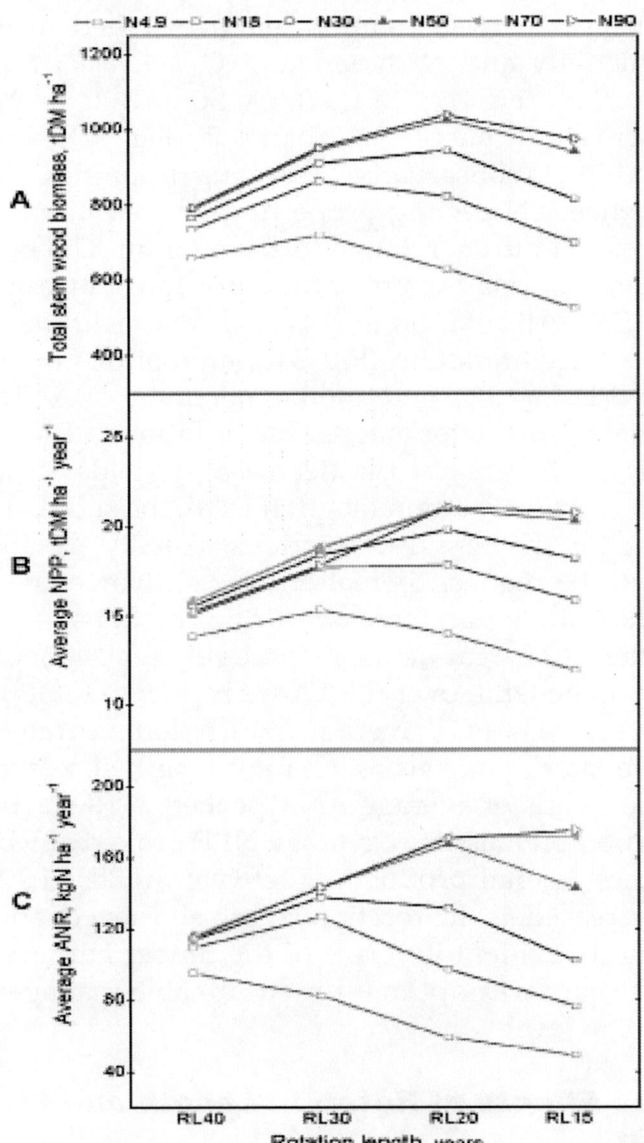

Figure 5. The long-term (total 120-year period) summary of interactive effects of increased N deposition and short-rotation management over successive rotations. It shows the sum of accumulative statues at the end-of-rotations, (A) stem wood biomass, and 120-year-average values for flux or rates, (B) NPP and (C) ANR.

doi:10.1371/journal.pone.0055376.g005

Over 120 year period, rotation length resulted in the changes in the productivity and N dynamic of *C. lanceolata* plantation with successive rotations (Fig. 3A, B, C, D, E). Soil available N decreased with the increase in successive rotations (Fig. 3D). First rotation plantation exhibited conservative N-cycling properties. As successive rotation practiced, N-cycling properties recovered with increasing availability of soil nitrate relative to ammonium. The dominance of a conservative N cycle for typical mature lowland tropical forests re-emerged [39]. All rotation lengths led to a yield decline in stem wood biomass accumulation (Fig. 3A), annual dry matter biomass increment (NPP, Fig. 3B), and annual increment of ANR (ANR, Fig. 3C) at harvest time over successive rotations. The shorter the rotation length, the greater the decrease in yields at harvest time (Figs. 3). For a certain rotation length (Fig. 4), the successive decrease was greatest for the first 1–3 rotations, usually by about 5–25%, and tended to be modest thereafter (Fig. 4). However, stem wood biomass demonstrated an increase during rotations 5–6 of the RL20 and 6–8 of the RL15 treatments, respectively (Fig. 4A). On the long-term (Fig. 5), in general, over the 120-year period, total productivity (total stem biomass and average NPP) demonstrated "increase first decrease later" patterns as rotation length shortened (Fig. 5A, B), while the trends of average ANR peaked at RL40 (Fig. 5C), and total stem wood biomass and average NPP peaked at RL30 (Fig. 5A, B). Differences in total production between RL30, RL20, and RL15 gradually increased as the rotation length shortened, and this result was particularly evident in terms of the timber harvest. A rotation length of 30 years was optimal for sustainable management at the N4.9 deposition level.

Interactive Effects of Rotation Length and Increased N Deposition

Rotation length clearly affected the productivity and N dynamics of the simulated *C. lanceolata*plantation over the 120-year period from perspective of growth response constraints (Figs. 3), successive effects (Fig. 4, Appendix S2), and long-term outcomes (Fig. 5, Appendix S2).

For different rotation lengths (RL40, RL30, RL20, and RL15) and the simulated productivity and N dynamic values of the *C. lanceolata* plantation over the 120-year cultivation period are presented in Fig. 3, 4 and 5. In general, on the long-term (Fig. 5), short rotation length had a more significant effect on productivity and N dynamics than high N deposition. Under all rotation length cycles, increasing N deposition generally led to increases in both productivity and N dynamics. However, the sensitivity of N deposition for a given rotation length decreased with increasing amounts of N deposition. Over the 120-year period, the total stem biomass, average NPP, and average ANR did not sensitively response to increase in N deposition level higher than N50 (50 kg N ha^{-1} $year^{-1}$). This result indicated that a N deposition level of 50 kg N $ha^{-1}year^{-1}$ was the upper limit level of thresholds for N deposition impacts for *C. lanceolata*plantations with rotation length of 30 and 40 years (Fig. 5). Similarly, N70 (70 kg N ha^{-1} $year^{-1}$) was the upper limit level of thresholds for N deposition impacts under a rotation length of 15 and 20 years (Fig. 5).

For all of the increased N deposition scenarios larger than 50 kg N ha^{-1} $year^{-1}$, all indicators increased as the rotation length was shortened from RL40 to RL20 over the entire 120-year simulated period and exhibited moderate variations in average ANR when the rotation length was changed from RL20 to RL15 (Fig. 5). These results indicate that the largest increase in total stem wood biomass and average NPP occurred when using the RL20 cycle and smaller differences in total stem wood biomass and average NPP occurred between RL15 and RL30 (Fig. 5A, B). At a specific N deposition level, over the 120-year period, a rotation length of 20–30 years would be the most sustainable practice in terms of total stem wood biomass, average NPP and average ANR under the N18 scenario (Fig. 5A, B, C), while under the N30 scenario, a rotation length of 20 years would be most sustainable in terms of total stem wood biomass and average NPP (Fig. 5A, B), and for N50, N70 and N90 scenarios, the sustainable rotation length would be 15–20 years in terms of total stem wood biomass and average NPP (Fig. 5A, B).

At the same time, a shorter rotation length resulted in a major decline in forest floor biomass. A change in the rotation length from RL40 to RL30 increased SOC by 0.15% under N18, 0.18% under N30, and 0.015% under N50, but decreased SOC by 15% under N70 and N90. Therefore, for N deposition levels higher than N18, rotation lengths <30 years lead to 2.5–8.2% losses in SOC (Fig. 3E).

Discussion

Uncertainties of Model Performance

The results of this study may be limited by four aspects of uncertainties: (1) resulted from complex interactions among the different components of the forest, varied in different forest regions, elevations, among different temporal and spatial scales by comprising multi-level of eco-biological complexity, the response of a forest ecosystem to climate change can have very different magnitudes and even different directions [40]–[42]. Available studies suggest that forest responses to climate change factors (such as increasing temperature and CO_2) will be limited by competition, disturbance, and nutrient limitations [43]–[44]. (2) CenW model was unable to acquire sufficient field measurements of reference parameters from literature, pertaining to model parameterization as it relates to *C. lanceolata* stands, fixed parameters of default values or average values from literature or field measurements were used in this study (see Appendix S1). (3) Our study concentrated on how N deposition and climate would affect the growth for *C. lanceolata* plantation. As mentioned before, we didn't consider land use change resulted from human activities, disturbances such as fires and environmental pollution as aerosols, atmospheric CO_2 and O_3, which are important factors affecting forest productivity and N dynamics [45]–[46]. (4) Although our results agree with many other studies at realistic temporal scales, but long-term productivity modeling still has limited knowledge of the forest responses to increased N deposition in the field on the long-term observation.

Model Simulation of Effects of Increased N Deposition

The simulated NPP of the *C. lanceolata* plantation demonstrated a steep increase during the first 10 years, reaching its maximum growth rate at the age of 13 years, which then declined thereafter. This growth pattern is in good agreement with the measured peak NPP values of evergreen coniferous forests in tropical and subtropical zones [47]. The simulated average NPP values during the first 60 years under the N4.9 scenario (4.9 kg N ha^{-1} $year^{-1}$) and RL120 length (11.34 t DM ha^{-1} $year^{-1}$) in this study are within the range of the average NPP (9.02–15.47 t DM ha^{-1} $year^{-1}$) measured in *C. lanceolata* plantations at the same site [48]. These results are close to those values measured in subtropical coniferous forests in China [49], [50]. Over the simulated 120-year period, the simulated average NPP (5.78 t DM ha^{-1} $year^{-1}$) is within the range of values measured in the forests of China (approximately 4.8–6.22 t DM $ha^{-1}year^{-1}$) [51], indicating that *C. lanceolata* is a fast growing tree species with a high production value. Despite some uncertainties existed in this study, the simulated results for initial N deposition level (N4.9) could be explained by the five factors reviewed by Ryan et al. [52], including changes in photosynthesis rate as stand development, nutrient supply, respiration, C allocation, and hydrological function. Our study also implied that a decline in stand leaf area usually accompanies a decline in aboveground wood growth. The decline in NPP with stand age can be explained as a combined effect of declining photosynthesis efficiency and declining N-availability for tree growth [53].

The results showed that of the productivity of *C. lanceolata* plantation in response to increased N deposition were more related to stand age than N addition depending on the proportion and age of growing forests. Increased N deposition has shown to raise the equilibrium state and constraint curve. Increased rates of N deposition can affect site nutrient stores and dynamics, microclimate, and tree growth. Recent evidence suggests that sustained growth response is unlikely to be realized as a cumulative effect of N fertilization, but such a possibility might exist with other forms of fertilization, if they become widespread [54]. Based on the assumption that duration of these effects is short, cumulative nutrient mineralization and/or leaching should be unlikely to occur as cumulative effects over time on single site. However, duplication of applications across landscape areas might hypothetically lead to cumulative increases in "baseline" nutrient mobilization and/or leaching for the area as a whole.

Interactive Effects of Increased N Deposition and Short-rotation Length

Maintaining long-term productivity in forests is a fundamental goal of ecologically sound forest management [55]. Factors that may be detrimental to the sustainability of plantations include the removal of a large proportion of material from the site, short-rotation length, and increased site disturbances [18]. Of particular concern is the sustainability of the physical properties of the soil, soil horizons, and complex biogeochemical cycles that are constantly at work within the forest system [28]. A number of studies have investigated the effects of harvesting regimes on the rotation cycle, growth rate, and C storage [18], [28], [32], [56]. Conventional stem-only harvesting is accepted as a sustainable forestry practice [57].

In this study, rotation length was an important factor that affects long-term productivity and N dynamics of *C. lanceolata* plantations if the N deposition level is low. As expected, total productivity over the 120-year simulation period demonstrated a nonlinear response pattern: specifically, an initial increase in productivity, which declined in accordance with a shorter rotation length. The highest value (15.25 t DM ha^{-1} $year^{-1}$) was achieved using a 30-year rotation cycle (RL30) (Fig. 5B). Therefore, under current climatic conditions and an N deposition level of 4.9 kg N ha^{-1} $year^{-1}$ (N4.9), a reasonable rotation cycle was determined to be 30 years in terms of sustainable productivity over the 120-year simulated period (Fig. 5B).

The rotation length also strongly affects the productivity and N dynamics over successive rotations at *C. lanceolata* plantations, which mostly monocultures that are currently managed using rotation lengths <20 years [28]. Short rotation cycles and intensive forest practices in the past have been linked to yield declines, soil depletion, and reductions in productivity over successive rotations in *C. lanceolata* plantations [4], [28], [30]. Significant declines in the yield (10–40%) have been reported [25], [27]. In this study, the simulated soil available N was highest in the second rotation for rotation lengths <40 years (RL40) (Fig. 3D). The simulated stem wood biomass, NPP, ANR, and SOC values were highest during the first rotation regardless of the rotation cycle, then, decreased as the number of rotation generations increased (Fig. 3A, B, C, E). The decreasing rate of SOC varied over successive rotations, demonstrating a sharp decrease by about 5–25% over the first 1–3 generations, but only a modest successive decrease was demonstrated thereafter. However stem wood biomass increased after the sixth RL20 rotation and the eighth RL15 rotation (Fig. 4A). The decline in yield over successive rotations, as observed in the model simulation, is consistent with the results of Zhang et al. [30], who reported that increments in the stand biomass were reduced by an average of 24% from the first to the second rotation and by a further 40% from the second to the third rotation. SOC was reduced by 10% between the first and the second rotations and by 15% between the second and the third rotations.

Changes in the composition and biomass of forest floors, associated with decomposition, nutrient mobilization, and redistribution from vegetation and forest floors to mineral soil horizons should occur commonly as a cumulative effect in managed forests, as these trends are influenced in similar or complementary ways by several forestry practices, including harvest, site preparation, fertilization and herbicide application. The long-term significance of these trends is unknown. Short-rotation management aimed to increase wood production may affect forest sustainability depending, among other factors, on initial site conditions and treatment regimes. Excessive removal of C and nutrients and soil disturbance are two of the major factors that may cause site degradation and reduction in forest productivity [58]. Short-rotation practices can increase productivity above base level and reduce or eliminate negative impacts. Currently, there is a little direct evidence to support that harvest removals in themselves lead to soil depletion over several succeeding rotations. Short-term productivity declines should be due to severe associated effects such as compaction, erosion or loss of organic layers.

The simulation results of this study indicate that the rotation length and N deposition level interactively affect the productivity and N dynamics of *C. lanceolata* plantations. Even though rotation length is an important factor that affects the productivity and N dynamics of *C. lanceolata* plantations, the offsite effects of the shorter rotation length decreased as the amount of N deposition increased. A high level of N deposition led to increased production and N accumulation in the plantation ecosystem, but the increasing effects of production due to high N deposition increased as the rotation length was shortened. These phenomena could be interpreted by the fact that high N deposition surpassed the loss of N in the soil due to the removal of harvesting materials and leaching due to runoff. Continuously increasing atmospheric N deposition will result in soil acidification, N leaching, nutrient imbalances, and NPP depression in forests [4], [59]. The results of this study imply that the offsite detrimental effects of chronic atmospheric N deposition could be ameliorated through appropriate forest management initiatives such as rotation length selection. For example, for a given level of N deposition, the responses of NPP to a high level of N deposition were strongly correlated with stand age (Fig. 2). Over the 120-year period, the total stem biomass, average NPP, and average ANR did not sensitively response to increase in N deposition level higher than N50 (50 kg N ha^{-1} $year^{-1}$). This result indicated that N deposition level of 50 kg N $ha^{-1}year^{-1}$ was the upper limit level of thresholds for nitrogen deposition impacts for *C. lanceolata*plantations with as rotation length of 30 and 40 years (Fig. 5). Similarly, N70 (70 kg N ha^{-1} $year^{-1}$) was the upper limit level of thresholds for nitrogen deposition impacts under a rotation length of 15 and 20 years (Fig. 5). The NPP sensitive to the N deposition level of 18–50 kg N ha^{-1} $year^{-1}$ (N18, N30 and N50) occurred when the stand age was 6–20 years, while the sensitive of NPP to high N deposition, such as N70 (70 kg N ha^{-1} $year^{-1}$) or N90 (90 kg N ha^{-1} $year^{-1}$), occurred when the stand age was 6–19 years. Therefore, the sensitive stand age of N increasing was around 20 years for atmospheric N deposition >18 kg N ha^{-1} $year^{-1}$ input to the *C. lanceolata*plantation. After the stand age of N sensitive, N deposition would exceed N uptake by the trees and the retention capacity, consequently resulting in negative environmental impacts. The sensitive stand age of N increasing for a specific N deposition level may be the appropriate rotation length to sustain productivity without offsite environmental effects in the *C. lanceolata*plantation.

The results of this study also indicate that increasing atmospheric N deposition might promote improvements in tree growth and site quality at *C. lanceolata* plantations. While very short rotation lengths reduced the capacity of the plantation ecosystem to accumulate C and N, particularly in sites with initially poor levels of nutrients, these reductions could have been averted by higher atmospheric N deposition. The obvious positive responses in terms of productivity and reduced N-use efficiency in *C. lanceolata* plantations due to increasing N deposition are consistent with another study on *C. lanceolata* plantations in southeastern China, which simulated results using the FORECAST model [32]. For example, increasing N deposition from 4.9 kg N kg N ha^{-1} $year^{-1}$ (N4.9) to 70 kg N ha^{-1} $year^{-1}$ (N70) resulted in higher productivity and timber biomass, increased forest N retention, and increased N requirements for tree growth, but lower SOC accumulation and decreased soil available N in *C. lanceolata*plantations. Further increasing N deposition to N90 (90 kg N ha^{-1} $year^{-1}$) or higher did not increase productivity, N retention, or affect SOC or soil available N. Therefore, N deposition>N70 (70 kg N ha^{-1} $year^{-1}$) leads to the loss of N through leaching and offsite effects.

CONCLUSIONS

Case studies that use modeling are needed to improve our understanding of the effects of the rotation cycle and N deposition on the long-term productivity of plantations. In terms of how the model simulations performed when they incorporated variables from a *C. lanceolata* plantation in subtropical China, our results show that CenW3.1 demonstrated considerable potential for this particular application. Not only can the model be used to integrate data from various forest studies in a dynamic manner, thus enabling a more thorough examination of study data, it can also be applied to predict the potential long-term effects of management practices and N deposition on the functions (such as C and N processes) of a forest ecosystem. Over the long term, a combination of indicators, including NPP, stem wood biomass and ANR indicate that 20–30 years is a sustainable rotation cycle for subtropical *C. lanceolata* plantations assuming an N deposition level of 4.9–50 kg N ha^{-1} $year^{-1}$; shorter rotation lengths (15–20 years) may be sustainable for N deposition rates between 50 and 70 kg N ha^{-1} $year^{-1}$, to avoid leaching and offsite effects.

Definitions, symbols, values, units, and sources of parameters used in CenW3.1 modeling of a *C. lanceolata* plantation. For the "Value sources" column, [8], [23], [31], [37], [65]-[72]are references citations, D = default, F = fitted, O = observed, and A = assumed.

ACKNOWLEDGMENTS

Special thanks to the CenW model group (Dr Miko U.F. Kirschbaum) for free access (http://www.kirschbaum.id.au/) to the CenW 3.1 modeling software.

AUTHOR CONTRIBUTIONS

Conceived and designed the experiments: WHX MFZ CHP. Performed the experiments: MFZ DLT XWD ZHH CHP. Analyzed the data: MFZ WHX XLZ. Contributed reagents/materials/analysis tools: CHP. Wrote the paper: MFZ WHX CHP.

REFERENCES

1. Heilman P, Norby R (1998) Nutrient cycling and fertility management in temperate short rotation forest systems. Biomass and Bioenergy 14(4): 361–370.
2. Moffat AS (1998) Global nitrogen overload problem grows critical. Science 279, 988–989.
3. Chen XY, Mulder J, Wang YH, Zhao DW, Xiang RJ (2004) Atmospheric deposition, mineralization and leaching of nitrogen in subtropical forested catchments, south China. Environ. Geochem. Health 26: 179–186. doi: 10.1023/b:egah.0000039580.79321.1a
4. Ma XQ, Heal KV, Liu AQ (2007) Nutrient cycling and distribution in different-aged plantations of Chinese fir in southern China. For. Ecol. Manag. 243(1): 61–75. doi: 10.1023/b:egah.0000039580.79321.1a
5. Lu XK, Mo JM, Gundersern P, Zhu WX, Zhou GY, et al. (2009) Effect of simulated N deposition on soil exchangeable cations in three forest types of subtropical China. Pedosphere 19(2): 189–198. doi: 10.1023/b:egah.0000039580.79321.1a
6. Ma XH (1989) Effects of rainfall on the nutrient cycling in man-made forests of*Cunninghamia lanceolata* and *Pinus massoniana*. Acta Ecologica Sinica 9: 15–20 (in Chinese).. doi: 10.1023/b:egah.0000039580.79321.1a
7. Zhou GY, Yan JH (2001) The influence of region atmospheric precipitation characteristics and its element inputs on the existence and development of Dinghushan forest ecosystems, Acta Ecologica Sinica. 21: 2002–2012 (in Chinese)..
8. Chen XY, Mulder J (2007) Atmospheric deposition of nitrogen at five subtropical forested sites in South China. Sci. Total Environ. 378: 317–330.
9. Hu ZY, Xu CK, Zhou LN, Sun BH, He YQ, et al. (2007) Contribution of atmospheric nitrogen compounds to N deposition in a broadleaf forest of southern China. Pedosphere 17(3): 360–365.
10. Fang YT, Zhu WX, Mo JM, Zhou GY, Gundersen P (2006) Dynamics of soil inorganic nitrogen and their responses to nitrogen additions in three subtropical forests, South China. J. Environ. Sci. 18: 756–763.
11. Maskell LC, Smart SM, Bullock JM, Thompson K, Stevens CJ (2010) Nitrogen deposition causes widespread loss of species richness in British habitats. Glob. Chang. Biol. 16: 671–679.

12. Zaehle S, Friedlingstein P, Friend AD (2010) Terrestrial nitrogen feedbacks may accelerate future climate change. Geophys. Res. Lett. 37: L01401. doi: 10.1029/2009gl041345
13. Clein JS, McGuire AD, Zhang X, Kicklighter DW, Melillo JM, et al. (2002) Historical and projected carbon balance of mature black spruce ecosystems across North America: The role of carbon-nitrogen interactions. Plant Soil 242(1): 15–32. doi: 10.1029/2009gl041345
14. Galloway JN, Aber JD, Erisman JW, Seitzinger SP, Howarth RW, et al. (2003) The nitrogen cascade. Bioscience 53: 341–356. doi: 10.1029/2009gl041345
15. Schoenholtz SH, Van Miegroet H, Burger JA (2000) A review of chemical and physical as indicators of forest soil quality: challenges and opportunities. For. Ecol. Manag. 138: 335–356. doi: 10.1029/2009gl041345
16. Hynynen J, Ahtikoski A, Siitonen J, Sievänen R, Liski J (2005) Applying the MOTTI simulator to analyse the effects of alternative management schedules on timber and non-timber production. For. Ecol. Manag. 207: 5–18. doi: 10.1029/2009gl041345
17. Kaipainen T, Liski J, Pussinen A, Karjalainen T (2004) Managing carbon sinks by changing rotation length in European forests. Environ. Sci. Policy 7: 205–219. doi: 10.1016/j.envsci.2004.03.001
18. Peng CH, Jiang H, Apps MJ, Zhang YL (2002) Effects of harvesting regimes on carbon and nitrogen dynamics of boreal forest in central Canada: a process model simulation. Ecol. Model. 155: 177–189. doi: 10.1016/j.envsci.2004.03.001
19. Pussinen A, Karjalainen T, Mäkipää R, Valsta L, Kellomäki S (2002) Forest carbon sequestration and harvests in Scots pine stand under different climate and nitrogen deposition scenarios. For. Ecol. Manag. 158: 103–115. doi: 10.1016/j.envsci.2004.03.001
20. Wei X, Kimmins JP, Zhou G (2003) Disturbances and the sustainability of long-term site productivity in lodgepole pine forests in the central interior of British Columbia: an ecosystem modeling approach. Ecol. Model. 164: 239–256. doi: 10.1016/j.envsci.2004.03.001
21. FAO (2006) Global planted forest thematic study: results and analysis. In: Lugo AD, Ball J, Carle J, editors. Planted Forests and Trees Working Paper. Rome. FP38E.
22. Lei JF (2005) Forest Resources in China. Beijing: China Forestry Publish House. 172 p. (in Chinese).
23. Wu ZL (1984) Chinese Fir. Beijing: China Forestry Publish House. (in Chinese).

24. Guo JF, Yang YS, Liu LZ, Zhao YC, Chen Z, et al. (2009) Effect of temperature on soil respiration in a Chinese fir forest. J. For. Res. 20: 49–53.

25. Tian DL, Xiang WH, Chen XY, Yan WD, Fang X, et al. (2011) A long-term evaluation of biomass production in first and second rotations of Chinese fir plantations at the same site. Forestry 84(4): 411–418.

26. Sheng WT, Fan SH (2005) Long Term Productivity of Chinese Fir Plantations. Beijing: Science Press. (in Chinese).

27. Ding YX, Chen JL (1995) Effect of continuous plantation of Chinese fir on soil fertility. Pedosphere 5: 57–66.

28. Bi J, Blanco JA, Seely B, Kimmins JP, Ding Y, et al. (2007) Yield decline in Chinese-fir plantations: a simulation investigation with implications for model complexity. Can. J. For. Res. 37(9): 1615–1630.

29. Zhang J, Wang SL, Feng ZW, Wang QK (2009) Stability of soil organic carbon changes in successive rotations of Chinese fir (*Cunninghamia lanceolata* (Lamb.) Hook) plantations. J. Environ. Sci. 21: 352–359.

30. Zhang XQ, Kirschbaum MUF, Hou ZH, Guo ZH (2004) Carbon stock changes in successive rotations of Chinese fir (*Cunninghamia lanceolata* (Lamb) hook) plantations. For. Ecol. Manag. 202: 131–147.

31. Zhao MF, Xiang WH, Peng CH, Tian DL (2009) Simulating age-related changes in carbon storage and allocation in a Chinese fir plantation growing in southern China using the 3-PG model. For. Ecol. Manag. 257: 1520–1531.

32. Wei XH, Blanco JA, Jiang H, Hamish Kimmins JP (2012) Effects of nitrogen deposition on carbon sequestration in Chinese fir forest ecosystems. Sci. Total Environ. 416: 351–361. doi: 10.1016/j.scitotenv.2011.11.087

33. Kirschbaum MUF (1999) Modelling forest growth and carbon storage with increasing CO_2 and temperature. Tellus B 51: 871–888. doi: 10.1016/j.scitotenv.2011.11.087

34. Kirschbaum MUF, Paul KI (2002) Modelling C and N dynamics in forest soils with a modified version of the CENTURY model. Soil Biol. Biochem. 34: 341–354. doi: 10.1016/j.scitotenv.2011.11.087

35. Running SW, Ramakrishna RN, Hungerford RD (1987) Extrapolation of synoptic meteorological data in mountainous terrain and its use for simulating forest evapotranspiration and photosynthesis. Can. J. For. Res. 17: 472–483. doi: 10.1016/j.scitotenv.2011.11.087

36. Nash JE, Sutcliffe JV (1970) River flow forecasting through conceptual models. J. Hydrol. 10: 282–290. doi: 10.1016/j.scitotenv.2011.11.087

37. Chen XY (1989) Dynamic characteristics of nitrogen in a Chinese fir plantation ecosystem. Journal of Central South Forestry College 9 (Sup.): 66–75. (in Chinese).

38. Yang YS, Guo JF, Chen GS, He ZM, Xie JS (2003) Effects of slash burning on nutrient removal and soil fertility in Chinese fir and evergreen broadleaved forests of mid-subtropical China. Pedosphere 13 (1), 87–96.

39. Davidson EA Reis de Carvalho (2007) Recuperation of nitrogen cycling in Amazonian forests following agricultural abandonment. Nature 447: 995–998www.nature.com/nature/journal/v447/n7147/full/nature05900.html.

40. Pastor J, Post WM (1988) Response of northern forests to CO2-induced climate change. Nature 334: 55–58. doi: 10.1038/334055a0

41. Overpeck JT, Rind D, Goldberg R (1990) Climate-induced changes in forest disturbance and vegetation. Nature 343: 51–53. doi: 10.1038/334055a0

42. Aber JD, McDowell W, Nadelhoffer K, Magill A, Berntson G, et al. (1998) Nitrogen saturation in northern forest ecosystems, hypotheses revisited. Bioscience 48: 921–934. doi: 10.2307/1313296

43. Galloway JN, Cowling EB (2002) Reactive nitrogen and the world: 200 years of change. Ambio 31: 64–71. doi: 10.2307/1313296

44. Davidson EA (2009) The contribution of manure and fertilizer nitrogen to atmospheric nitrous oxide since 1860. Nat. Geosci. 2: 659–662. doi: 10.2307/1313296

45. Ren W, Tian H, Tao B, Chappelka A, Sun G, et al. (2011) Impacts of tropospheric ozone and climate change on net primary productivity and net carbon exchange of China's forest ecosystems. Glob. Ecol. Biogeogr 20: 391–406. doi: 10.2307/1313296

46. Sitch S, Cox PM, Collins WJ, Huntingford C (2007) Indirect radiative forcing of climate change through ozone effects on the land-carbon sink. Nature 448: 791–794. doi: 10.1038/nature06059

47. Wang SQ, Zhou L, Chen JM, Ju WM, Feng XF, et al. (2011) Relationships between net primary productivity and stand age for several forest types and their influence on China's carbon balance. J. Environ. Manag. 92(6): 1651–1662. doi: 10.1016/j.jenvman.2011.01.024

48. Tian DL, Pan WC, Zhang CJ, Lei ZX (1993) The Characteristics of Biomass in Chinese Fir Plantation Ecosystem. In: Liu XZ, editor. Long-term Located Research on Forest Ecosystem. Beijing: China Forestry Publish House. 221–227 p. (in Chinese).

49. Wen DZ, Wei P, Kong GH (1999) The productivity and turnover of fine roots of southern subtropical forest in Dinghu Mountains. Acta Botany Sinica 23: 361–369 (in Chinese).. doi: 10.1016/j.jenvman.2011.01.024

50. Yang YS, Chen GS, He ZM, Chen YX, Guo JF (2002) Production, distribution and nutrient return of fine roots in a mixed and a pure forest in subtropical China. Chinese Journal of Applied Environment and Biology 8: 2230–233 (in Chinese).. doi: 10.1016/j.jenvman.2011.01.024

51. Fang JY, Liu GH, Xu SL (1996) Biomass and net production of forest vegetation in China. Acta Ecologica Sinica 16 (5): 497–508 (in Chinese).. doi: 10.1016/j.jenvman.2011.01.024

52. Ryan M G (1991) The effects of climate change on plant respiration. Ecol. Appl. 1: 157–167. doi: 10.2307/1941808

53. Murty D, McMurtrie RE (2000) The decline of forest productivity as stands age: a model-based method for analyzing causes for the decline. Ecol. Model. 134(2/3): 185–205. doi: 10.2307/1941808

54. Morford SL, Houlton BZ, Dahlgren RA (2011) Increased forest ecosystem carbon and nitrogen storage from nitrogen rich bedrock. Nature 477: 78–81. doi: 10.1038/nature10415

55. Helms JA (1998) Dictionary of forestry, Society of American Foresters and CABI Publishing. 210 p.

56. Xin ZH, Jiang H, Jie CY, Wei XH, Blanco J, et al. (2011) Simulated nitrogen dynamics for a *Cunninghamia lanceolata* plantation with selected rotation ages. Journal of Zhejiang Agriculture and Forestry University 28(6): 855–862 (in Chinese).. doi: 10.2307/1941808

57. Hakkila P (2002) Operations with reduced environmental impact. In: Richardson J, Bjorheden R, Hakkila P, Lowe AT, Smith CT, editors. Bioenergy from sustainable forestry: Guiding principles and practice. Dordrecht: Klewer Academic Publishers. 244–261 p.

58. Miller RE, Colbert SR, Morris LA (2004) Effects of heavy equipment on physical properties of soils and on long-term productivity. A review of literature and current research. National Council for Air and Stream Improvement Tech. Bull. No. 887. Research Triangle Park, NC. 76 p.

59. Weetman GF, McWiliiams ERG, Thompson WA (1992) Nutrient management of Douglas-fir and western hemlock stands: The issues. In: Chappell HN, Weetman GF, Miller RE, editors. Forest Fertilization: Sustaining and improving nutrition and growth of western forests. Institute of Forest Resources, College of Forest Resources, University of Washington, Seattle, WA, Contribution No. 73. 17–27 p.

60. Fang X, Tian DL, Cai BY, Gao YM (2002) Absorption, accumulation and transportation of N, P, K elements in the second-generation

Chinese fir plantation. Journal of Central South Forestry University 22(2): 1–6 (in Chinese)..

61. Tian DL, Kang WX (1993) Study on the Chinese Fir Plantation Adjustment Function to Water Cycle. In: Liu XZ, editor. Long-term Located Research on Forest Ecosystem. Beijing: China Forestry Publish House. 203–208 p. (in Chinese).

62. Wen SZ, He BF (1993) Studies on Runoff Law in Different interference of Chinese Fir Plantation Ecosystem. In: Liu XZ, editor. Long-term Located Research on Forest Ecosystem. Beijing: China Forestry Publish House. 221–227 p. (in Chinese).

63. Tian DL, Xiang WH (1993) Study on Dynamics Regularity of Soil Moisture in Chinese Fir Plantation. In: Liu XZ, editor. Long-term Located Research on Forest Ecosystem. Beijing: China Forestry Publish House. 209–215 p. (in Chinese).

64. Chen R, Xiang WH, Xu X, Tian DL, Liu J, et al. (2010) Speed of soil nitrogen mineralization and effect of fertilizer combined with warming on it in different ages Chinese fir plantations. Hunan Agricultural Sciences 5: 125–129 (in Chinese)..

65. Tian DL, Kang WX, Chen XY, Wen SZ, Wei XM (1989) Study on Microclimatic Characteristics of Chinese Fir Plantation Ecosystem in Small Watersheds. Journal of Central South Forestry College 9(Sup): 29–37 (in Chinese)..

66. Xue L, He YJ, Qu M, Xu Y (2005) Water holding characteristics of litter in plantations in south China. Chinese Journal of Plant Ecology 29(3): 415–421 (in Chinese)..

67. Kang WX, Zhao ZH, Deng XW (2007) Study of the Dynamic Effects and the Law of Kinetic Energy Transmission in the Canopy of Chinese fir Plantation Ecosystems. Journal of Central South University of Forestry & Technology 27(2): 1–6 (in Chinese)..

68. Yin H (2009) Improvement of Crown Interception Model Based on Vegetation Structural Parameters. Beijing: Beijing Forestry University.

69. Pan WC, Tian DL, Chen XY, Wen SZ (1989) Hydrological process and nutrient dynamics of a subtropical Chinese fir plantation ecosystem. Journal of Central South Forestry College 9 (Sup.): 1–10. (in Chinese).

70. Gu FX, Yu GR, Wen XF, Tao B, Li KR, et al.. (2008). Drought effects on carbon exchange in a subtropical coniferous plantation in China. Chinese Journal of Plant Ecology 32(5), 1041–1051. (in Chinese).

71. Qian N (1989) The relationship of leaf and foliage biomass to sapwood area in Chinese Fir. Journal of NanJing Forestry University 13(4): 75–80 (in Chinese)..

72. Liu XZ, Kang WX, Wen SZ (1993) Studies on energy balance on canopy in a Chinese fir plantation. In: Liu XZ, editor. Long-term Located Research on Forest Ecosystem. Beijing: China Forestry Publish House. 221–227. (in Chinese).

Chapter 2

CROP EXPANSION AND CONSERVATION PRIORITIES IN TROPICAL COUNTRIES

[1]Ben Phalan, [2]Monika Bertzky, [3]Stuart H. M. Butchart, [4]Paul F. Donald, [2,5]Jörn P. W. Scharlemann, [3]Alison J. Stattersfield, [1]Andrew Balmford

Ben Phalan, Andrew Balmford

[1]Department of Zoology, University of Cambridge, Cambridge, United Kingdom

[2]United Nations Environment Programme World Conservation Monitoring Centre, Cambridge, United Kingdom

[3]BirdLife International, Cambridge, United Kingdom

[4]Royal Society for the Protection of Birds, Sandy, United Kingdom

[5]School of Life Sciences, University of Sussex, Brighton, United Kingdom

ABSTRACT

Expansion of cropland in tropical countries is one of the principal causes of biodiversity loss, and threatens to undermine progress towards meeting the Aichi Biodiversity Targets. To understand this threat better, we analysed data on crop distribution and expansion in 128 tropical countries, assessed changes in area of the main crops and mapped overlaps between conservation priorities and cultivation potential. Rice was the single crop grown over the largest area, especially in tropical forest biomes. Cropland in tropical countries expanded by c. 48,000 km^2 per year from 1999–2008. The countries which added the greatest area of new cropland were Nigeria, Indonesia, Ethiopia, Sudan and Brazil. Soybeans and maize are the crops which expanded most in absolute area. Other crops with large increases included rice, sorghum, oil palm, beans, sugar cane, cow peas, wheat and cassava. Areas of high cultivation potential—while bearing in mind that political and socio-economic conditions can be as influential as biophysical ones—may be vulnerable to conversion in the future. These include some priority areas for biodiversity conservation in tropical countries (e.g., Frontier Forests and High Biodiversity Wilderness Areas), which have previously been identified as having 'low vulnerability', in particular in central Africa and northern Australia. There are also many other smaller areas which are important for biodiversity and which have high cultivation potential (e.g., in the fringes of the Amazon basin, in the Paraguayan *Chaco*, and in the savanna woodlands of the Sahel and East Africa). We highlight the urgent need for more effective sustainability standards and policies addressing both production and consumption of tropical commodities, including robust land-use planning in agricultural frontiers, establishment of new protected areas or REDD+ projects in places agriculture has not yet reached, and reduction or elimination of incentives for land-demanding bioenergy feedstocks.

INTRODUCTION

Cropland expansion as a threat to biodiversity. No human activity has altered the face of the planet more than agriculture [1]–[3]. Cropland covers at least 12% of the planet's ice-free surface, and annually we now harvest more than 10% of the Earth's net primary production in the form of crops [4], [5]. Although some species can benefit from agriculture [6], habitat loss resulting from its expansion is one of the greatest global threats to biodiversity [7]–[10] and threatens to undermine progress towards meeting the Aichi Biodiversity Targets [11]. Despite this, there have been few attempts to summarise and synthesise information on global patterns of crop expansion or cultivation potential in relation to priority areas for biodiversity conservation, or to carry out systematic assessments to identify which crops might pose the greatest threat to biodiversity [10], [12], [13].

Increases in food production in recent years owe more to intensification of crop production than to cropland expansion [14]. Projections suggest that land expansion will account for only 20% of production increases in developing countries in coming decades, with higher yields (including through increased multiple cropping and shorter fallow periods) accounting for the rest [14],[15]. But despite its modest contribution to global food production, meeting 20% of production increase from new cropland by 2030 would require conversion to crop production of an area equivalent to South Africa. Most of this land is likely to be in sub-Saharan Africa and South America [16]. Although the rate of global cropland expansion is slowing, there is little room for conservationists to be complacent: new croplands have in recent decades come largely at the expense of natural habitats, particularly tropical forests [17], [18]. New markets such as those for liquid biofuels are creating new demand for agricultural products [19]. The net effects on biodiversity of increased biofuel production depends on whether biofuels ameliorate climate change impacts sufficiently to offset their land-use impacts. If even a small proportion of crop-based biofuels are planted on previously carbon-rich land, or cause indirect land use change onto such land, biofuels overall will not help to reduce greenhouse gas emissions, at least in the near term [20]. Proposals for reducing emissions from deforestation and forest degradation (REDD) for climate mitigation might help to slow cropland expansion into forests, but there is also a risk that they will displace expansion into non-forest biomes [21]. As long as agricultural expansion continues, it seems likely to remain a major driver of biodiversity loss.

It is necessary to identify those crops that have expanded most rapidly in recent years (both in absolute and relative terms) and to assess the spatial pattern of these changes—especially in tropical biomes where most species occur—if we are to understand the current and future threats they pose. Individual crops differ enormously in their biodiversity impacts, depending on how and where they are cultivated [10]. Likewise, the drivers of expansion differ among crops, depending on socioeconomic context (e.g., whether demand is for subsistence use or overseas markets) and end uses (e.g., food, animal feed or biofuels) [22]. The impacts of crop cultivation also depend on the extent to which croplands are integrated into mosaics with natural and semi-natural habitats, in which case they might cause fragmentation over a wide area but have higher biodiversity value at a local scale; or are concentrated on a smaller total area, in which case they might have lower biodiversity value locally but affect a smaller area overall [8], [23], [24].

AIMS OF THIS STUDY

The aim of this paper is to provide a global overview of patterns of crop expansion in relation to conservation priorities in tropical countries. Specifically, we address the following questions:

1. Which crops cover most area in tropical countries and tropical biomes?
2. In which tropical countries has most expansion occurred in recent years, and which crops were involved?
3. How are remaining areas of cultivation potential distributed across tropical countries, particularly in relation to priority areas for biodiversity conservation?

We focus on tropical countries because they support the highest concentrations of species richness and endemism for most well-studied taxonomic groups, have large projected increases in demand for food from human populations growing in size and wealth, are experiencing high rates of habitat loss, and are seen as providing the most scope for increasing global agricultural production [14], [25]–[28]. An understanding of patterns of crop expansion across tropical countries in particular will therefore be essential if increasing conflicts between biodiversity conservation and human demands for agricultural products are to be addressed.

METHODS

Geographic Scope

Tropical Countries

We defined tropical countries as those with at least one-third of their land area between the Tropics of Cancer and Capricorn. This included 128 tropical countries (see Table S2). We used this definition rather than a wider definition incorporating all countries with any land in the tropics because pan-tropical data on changes in the area of specific crops were only available on a whole-country level; our definition thus excludes countries such as China and the United States, which have almost all of their territory outside the tropics.

Tropical Biomes

We clipped a global map of biomes [29] to the extent of tropical countries. Biomes included in analyses were (with shortened names used on figures in parentheses): 'tropical & subtropical moist broadleaf forests' (moist broadleaf forests), 'tropical & subtropical grasslands, savannas and shrubland' (grasslands, savannas), 'tropical & subtropical dry broadleaf forests' (dry broadleaf forests), 'deserts and xeric shrublands' (drylands), 'tropical & subtropical coniferous forests' (coniferous forests), 'montane grasslands and shrublands' (montane grasslands), 'mangroves' (mangroves) and 'flooded grasslands and savannas' (flooded grasslands). We excluded from the analyses all exclusively temperate or mediterranean biomes, and also lakes, rock and ice, and tundra.

Priority Areas for Biodiversity Conservation

We obtained GIS datasets of priority areas for biodiversity conservation as summarised by Brooks et al. [30] from various sources ([31]–[39] and see Table S4). We converted these, in a WGS84 geographic projection, to a 5 min×5 min (≈10 km×10 km) grid to match crop datasets, including any 5-min grid cell which overlapped the priority areas. These data were then imported into a PostgreSQL database. We used SQL queries to calculate areas of overlap using data from a matching grid on cropland extent and cultivation potential, and on the area of each grid cell calculated in an equal-area Behrmann projection.

Data Sources and Limitations

To explore the impact of different crops on priority areas for biodiversity conservation across the tropics we needed data on where they are grown and expanding, but available data vary in resolution and quality. Several land cover maps show global croplands, but they often use different definitions, with often quite different results [40]–[43]. Maps which integrate satellite-derived land cover data with subnational agricultural inventory data are probably more accurate[5], and now include global maps of individual crops [44], [45]. However, time-series of such maps are not yet available, so attempts to assess change are limited to using annual data at country level [46]. We use two sorts of such data: crop data (harvested area) for changes in area of individual crops, and land data (not differentiated by crop) for changes in cropland area[46]. (See Table S4 for further details of data sources.)

Analyses based on these global data must be interpreted critically, because their quality and consistency vary [47]. Three examples serve to illustrate the need for caution when interpreting such data:

1. India does not report any harvested area for oil palm fruit in FAOSTAT [46], although it has up to 1,780 km^2 of oil palm plantations [48]. If this was all harvested area, it would put India in the top 10 countries globally for oil palm area.
2. The crop responsible for most deforestation in Colombia, coca, is illegal and thus cannot be included in official FAO statistics [49].
3. Particularly in many African countries, crops are often intercropped on the same land [50]. The FAO provides advice for evaluating and reporting their area [51], but doing so consistently and accurately is inevitably difficult.

There are several further reasons why the sum of crop data might not equal that of land data. First, land where annual crops are harvested more than once per year from a given area is double- or triple-counted in crop data, but counted only once in land data [14]. Double- and triple-crop rice systems in Asia account for about 25% of global rice production [52]. Second, crop data exclude areas not harvested because crops were destroyed by drought, flooding or pests, or temporarily fallow, whereas land data typically include such areas [53]. Third, some countries report only fruit-bearing area for perennial crops, while others report all planted area[54]. The first discrepancy will cause crop data to overestimate true cropland area, while the second and third will lead to underestimates. Land data may thus give a more accurate picture of overall changes in cropland area.

Maps of cultivation potential [55] must also be interpreted with caution. First, climate data and projections are downscaled from a coarser grid [56]. Second, it is difficult to predict how technologies such as crop breeding will affect agricultural potential in the future: the dramatic expansion of soybeans in the southern Brazilian Amazon [57] for example, has relied upon the development of aluminium- and low-calcium-tolerant varieties [12], [58]. Third, social and political factors are important: the disastrous Mega Rice Project in Kalimantan is an example of politically-motivated cropland expansion in an area poorly suited to rice cultivation [59]. Nevertheless, while cultivation potential is not the only factor that will affect future patterns of crop expansion, and might not be the most important factor, mapping it helps to give a broad indication of the areas that might be vulnerable to conversion in the future.

To assess the possible impacts of crop expansion on biodiversity, we compared crop maps with priority areas for biodiversity conservation [30]. Brooks et al. [30] classified nine priority templates along axes of "vulnerability" and irreplaceability, defining "vulnerable" areas as those with little remaining habitat (high levels of past habitat loss). The definition is therefore retrospective, and does not provide information on vulnerability to threats in the future. Recent analysis using global land-use change projections from the IMAGE model [60] has suggested that some of the areas identified by Brooks et al. as being of "low vulnerability"—particularly High Biodiversity Wilderness Areas—might be highly vulnerable to agricultural expansion in the coming century [61].

Our analyses did not consider other forms of land use, such as livestock grazing, forestry and residential and commercial development. Conversion to cattle pasture remains the dominant driver of deforestation in Latin America, where over three-fifths of recent global humid forest conversion has occurred [62], [63]. We focus on cropland expansion because it changes habitat structure so profoundly, can be more accurately assessed by remote sensing (compared to many forms of grazing and forestry) and is so extensive (compared to urban areas).

Cropland Extent

Cropland Extent by Country: Crop Data

We extracted data on the harvested area of all crops for all 128 tropical countries for the years 1999–2008, the most recent for which data exist, from FAOSTAT [46]. We summed harvested areas of each of these 146 crops in each year to produce estimates of total harvested area for each crop. We also classified crops as annual or perennial [64], and summed areas of each of these two classes for each country in each year. Crops that can be grown as either annuals or perennials were classified according to [64]. For example, cassava, cotton and sugar cane were classed as annual crops (see Table S3 for scientific names of crops).

Cropland Extent by Country: Land Data

For each tropical country, we extracted data for 1999–2008 on the area of 'arable land', which corresponds to the area occupied by annual crops, and of 'permanent crops' (which in turn corresponds to the area occupied by perennial crops [53]). These data are reported in aggregate, without information on specific crops.

Maps of Tropical Cropland Extent

We obtained maps showing the spatial distribution of cropland [5] and of individual crops [44]. Each map shows the percentage of cropland (or of specific crops) per 5-min (≈10 km) grid cell. Other similar datasets exist [45], but the maps we used were the only ones which integrated satellite and detailed subnational inventory data, and which included all of the major tropical crops. Smaller island groups, including several of high biodiversity value such as Hawai'i, the Galápagos, the Solomons, New Caledonia and Fiji, do not feature on these maps. These were included in crop and land data totals (see above), but excluded from spatial analyses.

Multiple Cropping

As explained above, harvested areas might in some cases overestimate actual land areas used for crops harvested more than once per year from a given area. We tested whether this would change our rankings for the 12 most important crops in our dataset (defined as those in the top 10 crops by harvested area in tropical countries, and/or the top 10 by annual area increment) by calculating the minimum harvested area for each of them using information on the distribution of multiple cropping zones. We first calculated the area of each crop grown within each of nine 'multiple cropping zones', using crop maps from Monfreda et al. [44] (which counts double-cropped areas twice) and cropping zones from plate 13 of Fischer et al. [65]. We then divided the harvested area found in each cropping zone by the number of harvests of that crop obtainable in a year in that zone (ranging from none to three) [65]. The sum of these smaller areas gave a minimum estimate of the actual area occupied by each crop. Analyses were carried out using a Behrmann equal-area projection in ArcGIS 9.3 [66]. Crop rasters in geographic projection (WGS84) were converted to polygons, and polygon-in-polygon analyses were used to calculate the proportion of each crop in each zone.

Cropland Extent by Tropical Biome

We estimated the proportion of each tropical biome occupied by cropland based on the map from [5] in ArcGIS, using similar methods to those described for multiple cropping (above). We did this in two ways. First, for each biome we calculated the mean proportion of land occupied by cropland, weighted by cell area, using information on the percentage of each 5-min grid cell occupied by cropland [5]. Second, we calculated the number and area of 5-min grid cells in each biome where there is cropland covering <10% and ≥10% of land. This second method better captures the extent of agricultural landscapes, roughly equivalent to the "villages" and "croplands" of [3], across tropical countries.

Crop Composition by Tropical Biome

We estimated the proportion of each tropical biome occupied by each of the 12 most important tropical crops, using the same method as for cropland. We calculated the proportion of each biome occupied by each crop, using crop maps for the year 2000 from [44].

Cropland Expansion

Individual Crops

To estimate the mean annual increment in harvested area of each crop across the tropics, we used linear regression of crop area on the years 1999 to 2008. We also calculated the minimum annual increment (taking account of multiple cropping) by adjusting the annual increment by our crop specific ratios of harvested area: minimum harvested area. We used regression to estimate annual change—rather than a simple comparison of area in 1999 with that in 2008—because using data points for each year (rather than just the start and end years) reduces the chance of inaccuracies in reporting having a large influence on trends, though we also looked at results based just on the difference in crop areas between 1999 and 2008.

Cropland Expansion by Country

We estimated mean annual increments for annual and perennial crops for each country, using both crop data and land data (see previous section). We used linear regression of cropland area on year to produce estimates of annual change for each country.

Cropland Potential

Mapping Cultivation Potential

To map the extent to which areas of highest cultivation potential are already occupied by cropland, we used maps of "agro-climatically attainable yield" for the 12 most important tropical crops [55]. Maps were averaged projections of yield over the period 2010 to 2030, based on a mid-range climate scenario (H_3B_2). We assumed an intermediate input level, except for crops mainly grown as cash crops (rice, wheat, soybeans, sugar cane and oil palm), where we assumed a high input level. For each 5-min grid cell, we could determine the potential yield for each crop, as a percentage of the tropical maximum for that crop. We took the value for the crop with the highest percentage in a grid cell as an indicator of cultivation potential, to produce combined maps of cultivation potential for crops with similar requirements for wetter climates (cassava, rice, sugar cane and oil palm) and for drier climates (beans, cow peas, groundnut, maize, millet, sorghum and soybeans), and of cultivation potential for the top 12 crops combined.

Cultivation Potential in Relation to Priority Areas for Biodiversity Conservation

We quantified the extent to which conservation priority areas in tropical countries are already occupied by cropland, and the extent to which the remaining land in these areas is suitable for rainfed crop production. We obtained shapefiles of the nine conservation priority templates presented in [30], clipped these to the extent of tropical countries, and converted them to a 5 min×5 min grid. We calculated an area-weighted mean of cropland extent within each template, using a map of cropland extent [5]. We then calculated the mean "cultivation potential" (as defined above) of the remaining land within each template, after subtracting land already converted to cropland. Grid cells for which cultivation potential was undefined, or that were located in water bodies, were not included in this calculation.

Results

Crop Data

The three crops with the greatest harvested area in tropical countries in 2008 (Figure 1, Table S1) were also those with the greatest harvested area globally: rice, maize and wheat [46]. The 10 most important crops by harvested area, which collectively make up two-thirds of all harvested area in tropical countries, also included sorghum, soybeans, millet, beans, sugar cane, cassava and groundnuts. All are annual crops. Rice was grown over the largest area in tropical countries (18% of tropical cropland), whereas wheat was grown over the largest area globally. When adjusted to take account of the potential for multiple cropping (minimum harvested area), the top 10 crops remained the same, although the order changed. Of the 146 crops for which data were available, 77 were annual crops and 69 were perennial crops.

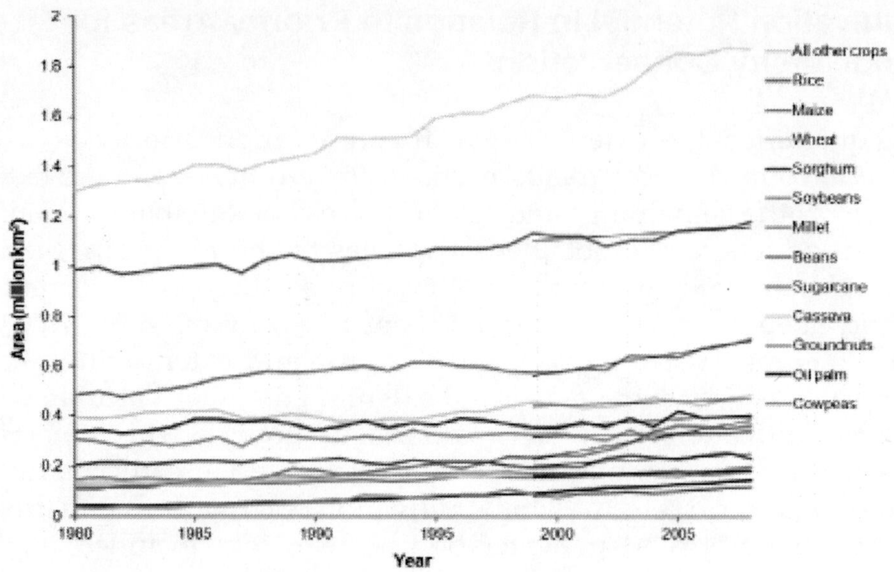

Figure 1. Harvested area of major crops in tropical countries, 1980–2008.The top ten crops in terms of their area in 2008 are shown. Oil palm and cow peas, which were the only two crops not on this list but which were in the top ten by area increase from 1999–2008, are also shown. Harvested areas of all other crops than these 12 are combined. Linear regressions used to assess recent rates of change in harvested area are shown. Source: [46].

doi:10.1371/journal.pone.0051759.g001

Land Data

The total area of cropland in tropical countries in 2008, calculated by summing the area of land used to grow both annual and perennial crops, was 6.7 million km^2. This was greater than the summed harvested area of all crops in 2008: 6.4 million km^2, suggesting that any overestimates of area introduced by multiple cropping were more than compensated for by underestimates caused by exclusion of unharvested cropland (as described in Data Sources and Limitations). To provide some context, 6.7 million km^2 is approximately twice the land area of India, or somewhat smaller than the land area of Australia.

Cropland Extent by Tropical Biome

Cropland made up 10.7% of the land area of tropical countries, a little less than the global figure of 12% of ice-free land [5]. It occupied 4–17% of the area of each biome, except for dry broadleaf forests which had 32% cropland cover (Figure 2A). Summing the total area of 5-min grid cells in which there was some cropland, 62–94% of each biome had some cropland, except drylands, where the figure was 22% (Figure 2B). Counting only grid cells with at least 10% cropland, 10–67% of each biome was occupied by agricultural landscapes.

Crop Composition by Tropical Biome

Rice was the most widespread crop in the moist broadleaf forests biome, followed by maize, wheat, soybeans, sugar cane and oil palm (Figure 3). In the grassland/savanna biome, sorghum, maize and millet dominated by area. In dry broadleaf forests, rice was again most widespread, followed by maize and soybeans. In drylands, wheat and millet were most widespread. Maize dominated in the coniferous forest biome and in montane grasslands. Rice was the main crop in the mangrove biome.

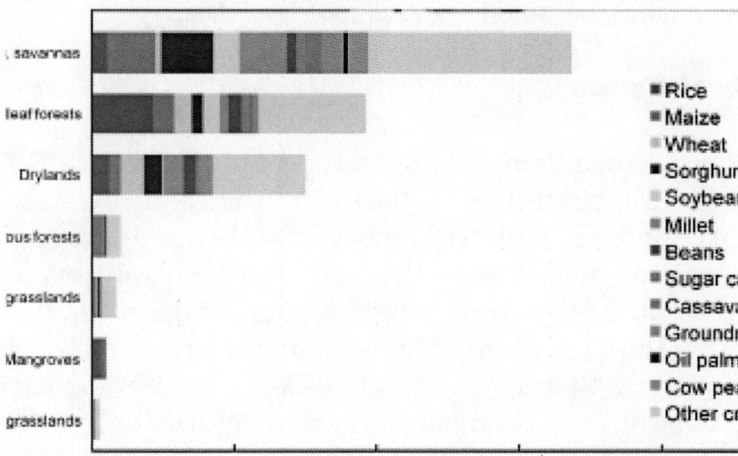

Figure 3. Area of different crops as a proportion of cropland in biomes within tropical countries. The top 12 tropical crops (see text) are identified.

The width of each bar in this figure is equivalent to the width of the brown portions of the bars in Figure 2A. Source: [44].

doi:10.1371/journal.pone.0051759.g003

Cropland Expansion

Overall

Across all tropical countries, cropland increased by on average ~48,000 km^2 per year, based on land data, or ~98,000 km^2 based on crop data. This equates to a rate of around 0.7% to 1.5% per year. Using only a simple change comparison between 1999 and 2008 (rather than regression models), these estimates were ~45,000 km^2 (land data) or ~86,000 km^2 (crop data) per year. As discussed in Data Sources and Limitations, the lower of each pair of estimates (based on land data) are likely to reflect more accurately land area converted to cropland, because multiple cropping and cropland from which crops were not harvested complicate reliable aggregation of area statistics from individual crops. Less than one-third of this increase (27.5%) was attributable to expansion of perennial crops (permanent crops), with the rest (72.5%) attributable to expansion of annual crops (arable land), based on regression of land data.

Individual Crops

In terms of the mean annual area added over the period 1999–2008, soybeans and maize were by far the two most rapidly expanding crops in tropical countries (Table S1). Only one of the top 10 was a perennial crop—oil palm—which was the fifth most rapidly expanding in harvested area, or the third when adjusted for multiple cropping. The 10 most important crops by area increment, which collectively account for more than two-thirds (69.7%) of the net increase in area in tropical countries, also included rice, sorghum, beans, sugar cane, cow peas, wheat and cassava. Eight crops were shared between both top 10 lists, while millet and groundnuts featured only in the top 10 by harvested area, and oil palm and cow peas only in the top 10 by annual area increment. Results were quite similar whether a simple change comparison or regression models were used: the order changed, but the identity of the top nine crops remained the same. The simple comparison produced estimates of change which were on average 8% smaller than those from the regression models.

Cropland Expansion by Country

Expansion of annual crops has occurred throughout most of the tropics (red circles in Figure 4A). Based on land data, the countries which added the greatest area of annual crops (absolute increase in arable cropland) over the period 1999–2008 were Nigeria, Sudan, Ethiopia, Brazil and Indonesia. These same five countries—in a different order—also experienced the greatest increases in cropland overall. The countries in which annual crops expanded at the greatest rate (relative to the area of cropland) were Sierra Leone, Guinea, Paraguay, Ethiopia and the Gambia (see Table S2 for further details). In several countries—including India, Australia, Colombia, Mexico and Thailand—the reported area of annual crops decreased. This could have been because of cropland degradation, or a genuine contraction of annual cropland because of conversion to other uses (including perennial crops) or increases in land-use efficiency. The magnitude of changes in the area of perennial crops in tropical countries (Figure 4B) was generally smaller than that of changes in area of annual crops. While cropland area (annual and perennial crops combined) expanded in 68 of 128 tropical countries, it declined in 40 others, and remained the same in 20 countries (almost all tiny island nations).

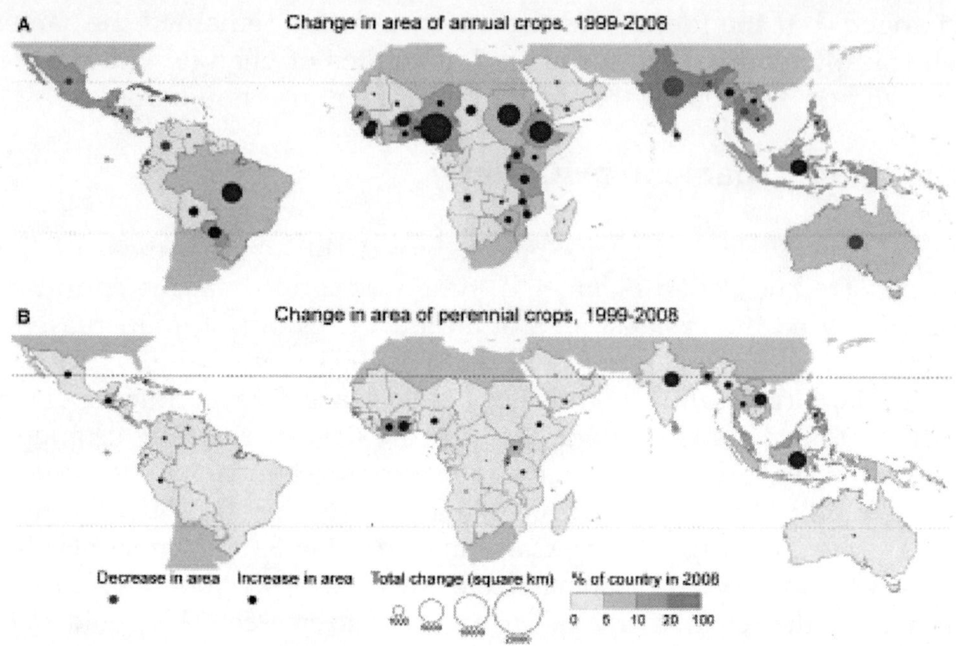

Figure 4. Increments in the area devoted to cropland in tropical countries. Circles show absolute increment over the period 1999–2008, with scale exaggerated 10 times for ease of interpretation. Shading indicates percentage of each country occupied by annual crops in 2008. Countries not defined as tropical are shaded grey. Maps are based on land data, for (A) arable land (annual crops) and (B) permanent cropland (perennial crops). Source: [46].

doi:10.1371/journal.pone.0051759.g004

Cropland Potential

Mapping Cultivation Potential

Most land in tropical countries, with the exception of deserts and high mountains, is suitable for crop cultivation (Figure 5). There are appreciable areas believed to have cultivation potential but with little or no cropland yet (mapped in dark blue in Figure 6), particularly in the fringes of the Amazon basin, across the Congo basin, and in northern Australia. Many other parts of the tropics that are most suitable for rainfed crop production are already heavily utilised for cropland (dark purple in Figure 5). Examples include large parts of Central America, the Caribbean, south-east Brazil, large parts of the African savannas, and much of south and south-east Asia, particularly the Sundaic lowlands. In a few places, crops are grown with the aid of irrigation on land with an otherwise unsuitable climate (red in Figure 6).

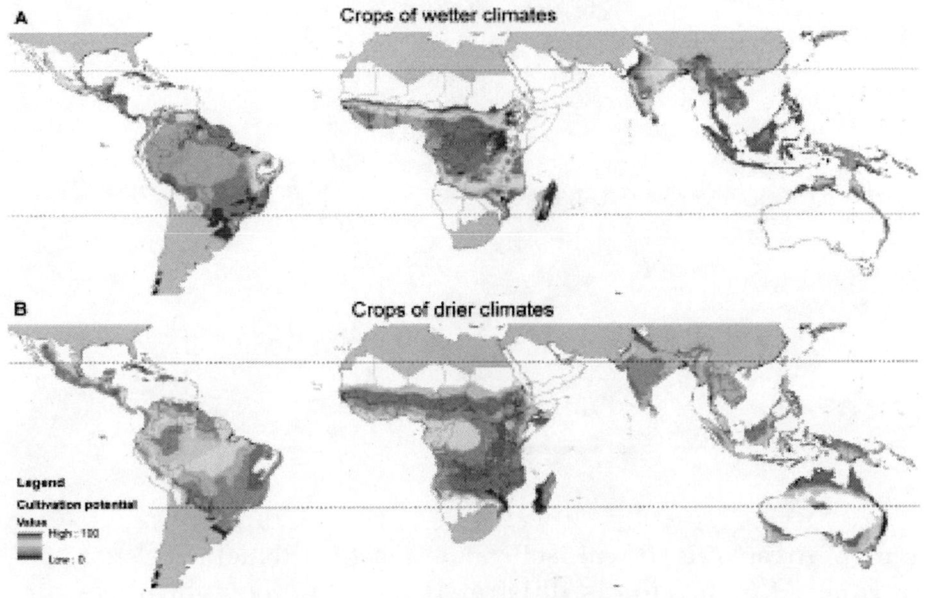

Figure 5. Areas of land with cultivation potential for selected crops of (A) wetter climates and (B) drier climates. Maps are based on four wetter-climate crops (cassava, rice, sugar cane and oil palm) and eight drier-climate crops (beans, cow peas, groundnut, maize, millet, sorghum, soybeans and wheat). The map shows cultivation potential for the crop for which each 5-min grid cell is most suitable. Cultivation potential is calculated as the "agro-climatically attainable yield" for each rainfed crop as a percentage of the global maximum for that crop [55].

doi:10.1371/journal.pone.0051759.g005

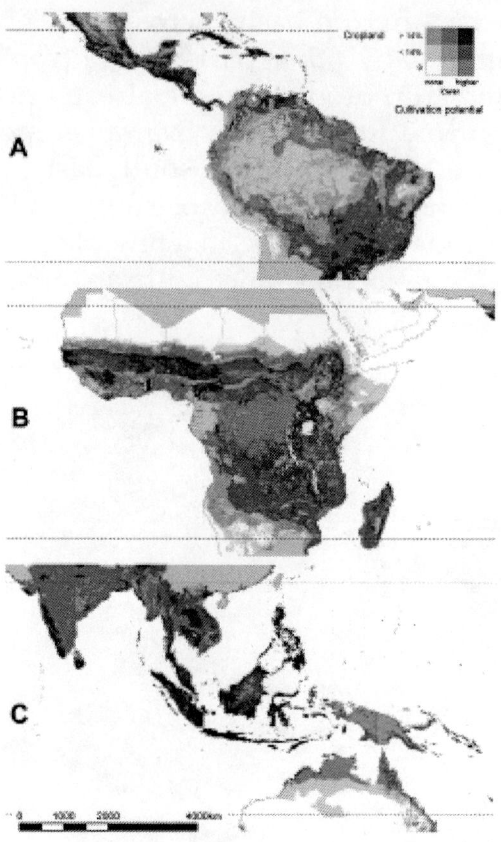

Figure 6. Areas of land with cultivation potential (blue) in relation to current cropland (red). This is illustrated for (A) Neotropical countries, (B) tropical Africa and (C) tropical Asia/Australia. Shades of blue indicate cultivation potential for the crop for which each 5-min grid cell is most suitable. Cultivation potential is calculated as the "agro-climatically attainable yield" for 12 major tropical crops as a percentage of the global maximum for that crop [55]. Shades of red indicate cropland extent in the year 2000, from [5]. The darker shades indicate values above the median. Land which is suitable for one or more crops, and which is already cultivated, is mapped in shades of purple. Land with no cultivation potential for these crops, and no cropland, is mapped in white, and land outside tropical countries is shaded grey.

doi:10.1371/journal.pone.0051759.g006

Cultivation Potential in Relation to Priority Areas for Biodiversity Conservation

As would be expected, those priority templates identified by [30] as having high (retrospective) "vulnerability" (Biodiversity Hotspots and Critical Ecoregions) were also those with the largest proportion of their area already converted to cropland (Figure 7). However, there was no clear relationship between retrospective "vulnerability" and future cultivation potential. Priority areas for biodiversity conservation previously identified as having low (retrospective) "vulnerability" included those with the lowest (Last of the Wild) and highest potential for future cultivation (Frontier Forests and High Biodiversity Wilderness Areas). It would appear therefore that retrospective assessments of "vulnerability" provide little information about whether areas are biophysically suitable for conversion to cropland in the future.

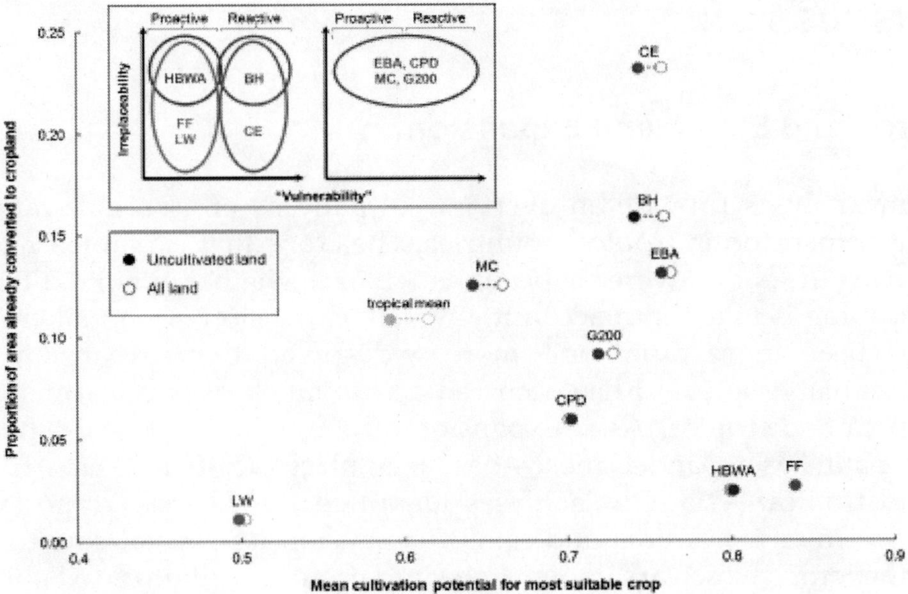

Figure 7. Cropland extent and cultivation potential within priority areas for biodiversity conservation in tropical countries. Cultivation potential is defined as in Figure 5. The open symbols show the mean cultivation potential of all land in each set of priority areas, while the filled symbols

show the mean cultivation potential of land that had not yet been converted to cropland as of 2000. Inset from [30] shows conservation priority templates placed within the conceptual framework of irreplaceability and (retrospective) "vulnerability" and coloured accordingly (reprinted with modification, with permission from AAAS). "Proactive" conservation priorities are those in areas which are not yet considered to be highly "vulnerable" to conversion, while "reactive" priorities are those in areas where there has already been much habitat conversion. Abbreviations: Biodiversity Hotspots (BH), Centres of Plant Diversity (CPD), Crisis Ecoregions (CE), Endemic Bird Areas (EBA), Frontier Forests (FF), Global 200 Ecoregions (G200), High Biodiversity Wilderness Areas (HBWA), Last of the Wild (LW), Megadiversity Countries (MC). Mean across all tropical countries shown by grey symbols.

doi:10.1371/journal.pone.0051759.g007

DISCUSSION

Cropland Extent and Expansion

Our analyses provide an overview of patterns of crop cultivation and expansion in tropical countries. The crops that expanded most during the period were soybeans and maize, whether or not multiple cropping is taken into account. Overall, expansion of annual crops has been more rapid and more widespread than expansion of perennial crops, and has occurred across much of South America, Africa and tropical Asia. Expansion of perennial crops—of which, oil palm has expanded most—has taken place mostly in West Africa and tropical Asia. Our analyses identified the 12 most important crops in terms of area and rate of expansion in tropical countries. Other crops which are known to be important contributors to habitat loss in specific places did not make it onto this list. Examples include cotton [67], coffee [68], tea [69], cocoa [70], rubber [71], coca [49] and pulp and paper [72].

Some of the crops which have expanded most in area in recent years are already well known drivers of biodiversity loss. Soybean expansion is recognised as a major cause of biodiversity loss in the Brazilian *Cerrado* savannas [57]. Oil palm has been described as 'the greatest immediate threat to biodiversity in Southeast Asia' [73]. Sugar cane has been implicated in the extinctions of species such as the Greater 'Amakihi *Hemignathus sagittirostris* in Hawai'i and the Alagoas Curassow *Mitu mitu* (Extinct in the Wild) in Brazil [74]. Expanding maize cultivation threatens the dry forests of Madagascar [75], [76], and rice cultivation is an important cause of wetland loss [10]. Others, such as sorghum, cow peas and millet have received much less attention in the conservation literature [77], although the combined area converted to these three crops in 1999–2008 was more than twice that converted to oil palm.

There are several possible reasons why some crops have received relatively little attention from conservationists. First, area is an incomplete proxy for impact. Coffee, for example, covers a relatively small area (8% of that occupied by rice in tropical countries), but tends to replace habitats of particularly high biodiversity value. Tropical drylands support unique species, but not concentrations of endemics on the scale of tropical forests, so hectare for hectare, dryland crops might have less of an impact on biodiversity than crops of wetter climates. Second, most of the less well-known crops are traditionally grown mainly by small-scale farmers rather than on an industrial scale by large corporations. As a consequence, there is less of a clear link to Western consumers [77]. Campaigns targeting the commodity supply chains of large corporations supplying European and North American retailers have been a key factor in efforts to reduce environmental impacts of commodities such as palm oil, coffee and cocoa [78]. However, even crops traditionally seen as the preserve of subsistence farmers are increasingly grown in large-scale commercial monocultures: sorghum in parts of the Caribbean and Latin America, and cassava in Thailand and Brazil, for example [12]. Such crops are also increasingly used for biofuels and animal feed rather than to feed people [12].

The Future of Wild Lands

One other, very recent study has suggested that High Biodiversity Wilderness Areas might be disproportionately affected by prospective patterns of cropland expansion in coming decades, as estimated by the IMAGE model [61]. Our analysis, based more directly on maps of cultivation potential, provides further evidence that not only High Biodiversity Wilderness Areas but also Frontier Forests have the biophysical attributes that could predispose them to future conversion (indeed, all nine sets of priority areas for biodiversity conservation have considerable cultivation potential, Figure 7). High Biodiversity Wilderness Areas and Frontier Forests have been considered to have "low vulnerability" because of low levels of past habitat loss, but are likely to come under threat as infrastructure develops and if political circumstances change. The Last of the Wild priority areas have much lower cultivation potential on average because they include large areas of desert [39]. Bearing in mind that our maps are probably inaccurate at a fine spatial scale, and that biophysical cultivation potential is just one of several determinants of vulnerability to agricultural conversion, the most extensive blocks of natural habitat in areas of high cultivation potential (dark blue in Figure 6) are in central Africa, the fringes of the Amazon Basin, and northern Australia.

Central Africa

The extent of land with cultivation potential in the Congo Basin in central Africa is particularly alarming. Most global conservation prioritisation schemes judge this area to be at low risk [30]because of its high forest cover and low recent rates of deforestation: 0.2% to 0.4% per year[79]–[81]. However the factors that may help explain the low rates of past deforestation – such as low population densities, low road density, political instability and lack of inputs to utilise poor soils [82], [83] – are changing [84]–[87]. As a result, the "last of the wild" in central Africa is increasingly fragmented [39], and there is "not much time" [38] to protect these forests from logging followed by conversion to cassava, oil palm, rice and sugar cane. There have been reports of large-scale land acquisitions for oil palm cultivation in the Democratic Republic of Congo, including one of 28,000 km^2 in 2007, but this appears to have been exaggerated and nothing has happened on the ground [88]. Nevertheless, the Congolian forests will continue to come under threat from expanding croplands, a threat which could be reduced by a strategic approach to road development [89], [90] and incentives for forest protection under a REDD+ mechanism [91].

Amazon Basin

The situation in the Amazon Basin is very different to that in central Africa. Recent deforestation has been more rapid and extensive, at least in Brazil, and thus has received far more attention from researchers and policy-makers [62], [92]. Almost half (46%) of the Brazilian Amazon has been formally protected within reserves, including indigenous reserves ([93], see Figure S1) and the rate of deforestation has declined in recent years [92]. However, legal protection for forests on private land is in danger of being weakened by changes to Brazilian legislation [94]. Our analysis confirms that most of the interior of the Amazon Basin is of relatively low suitability for agriculture, albeit still with similar cultivation potential to large parts of India or West Africa (Figure 5). Land with higher cultivation potential is concentrated around the fringes of the Amazon Basin. In unprotected areas with cultivation potential, e.g., in the Guiana Shield, creating new protected areas and setting limits on road expansion could help to reduce the threat from crop expansion [95].

Northern Australia

Multiple attempts to establish crops in the "empty north" of Australia have had limited success, because of intense seasonality and poor, easily eroded soils [96]. Agriculture in that area is dominated by cattle farming rather than crop production. However, minimal or no-tillage systems can enable integration of cropping and grazing, and with declining rainfall elsewhere on the continent, interest is again shifting to northern Australia. Aboriginal land and protected areas cover large parts of the suitable area mapped in Figure 6, but the risk of cropland expansion in unprotected land merits concern.

Other Parts of the World

Other, smaller areas of high cultivation potential but as yet with little cropland are mapped in dark blue in Figure 6, and include parts of the Paraguayan *Chaco* and the savanna woodlands in the Sahel and East Africa. Areas which are suitable and already heavily farmed (dark purple) include moist and dry tropical forests in coastal Mexico and Cuba, moist tropical forests and savanna in southeastern Brazil (*Mata Atlântica* and *Cerrado*), much of West Africa (especially Nigeria), Uganda, parts of India, and much of South-east Asia. Large tracts of South America, Africa and Southeast Asia are suitable for cropland but are not yet heavily farmed (bluish purple): instead, these areas are a mosaic of croplands and fragments of semi-natural or natural habitats [3].

Reactive or Proactive Conservation?

Brooks et al. [30] classify priority areas for biodiversity conservation as being either "reactive" or "proactive". Reactive areas are those with "high vulnerability", and include Biodiversity Hotspots and Crisis Ecoregions. Proactive areas are those with "low vulnerability", and include High Biodiversity Wilderness Areas, Frontier Forests and the Last of the Wild. As Brooks et al. acknowledge, the measures of vulnerability that were used to identify these areas relied mainly on past patterns of habitat loss, and made no effort to be predictive. It is perhaps unsurprising, then, that we found no consistent relationship between "vulnerability" *sensu* Brooks et al., and cultivation potential (Figure 7) which is one component of vulnerability to future conversion.

What are the implications of this observation for conservation priorities? It means that much of the land which has not yet been converted to farmland has not been left alone because it is uncultivable, but because political or socio-economic factors have impeded, or at least not promoted, conversion so far. With global demand for land rising [97] areas that could previously safely be considered to have "low vulnerability" may come under increasing threat from agricultural expansion. In addition to working in areas of "high vulnerability", conservation organisations might therefore be wise to increase their proactive conservation efforts while substantial opportunities for conservation in areas of "low vulnerability" still exist.

Reliability of the Maps and Data Used

The conclusions discussed above hinge on the reliability of the data used, and in particular the reliability of the maps of cultivation potential for tropical crops. As discussed in the Methods, there are good reasons to interpret these with caution. The discrepancy between our two estimates of cropland area in tropical countries is equivalent to the area of Italy, or to six years' worth of cropland expansion. For the reasons discussed in Data Sources and Limitations, we think the larger figure of 6.7 million km^2 is likely to be closer to the truth, but this cannot be independently verified.

In relation to the maps, the soil, terrain and climate datasets used to produce them are themselves coarse-grained and have not been comprehensively ground-truthed. Global datasets should not be taken to provide a detailed picture of conditions at finer scales. The maps do not take into account irrigation: where aquifers or rivers exist, this can transform land without sufficient rainfall for agriculture.

The maps also cannot capture new developments in agricultural technology, which allow crops to be grown where previously they could not be. A striking example is the *Cerrado* of Brazil. Until recent decades this savanna area was considered "unfit for farming" [98]. However, this is the new agricultural frontier of Latin America, where the annual area deforested for agriculture (pasture and cropland) is now on a par with that in the Amazon [99] and which is experiencing rapid expansion of cash crops such as soybeans and cotton [100], [101]. This is a clear example of how areas which are unsuitable today may become suitable in the future through developments in technology.

A key research need is therefore to reduce the uncertainties in these maps and to develop credible, fine-grained maps of cultivation potential which can be used in strategic planning, both to ensure that crops are grown where they will be most productive, and so that threats to biodiversity from agriculture can be better understood and avoided. For some crops and countries fine-grained maps have been developed (for example [102], [103]), but as they typically have not been comprehensively ground-truthed (if at all) it is difficult to know whether they are any more reliable than global datasets.

Beyond the need for better maps of cultivation potential, there is also a need for better models of future land-use change. On its own, cultivation potential is a relatively poor predictor of conversion risk. As our results here show, there are parts of the world with high cultivation potential that are not farmed, and other places with low cultivation potential which are. Other factors such as accessibility, socio-economic conditions, land tenure and government policies have as much, if not more, influence on where land conversion takes place. These drivers and policies differ greatly between countries and regions, and therefore the most promising way forward for anticipating future cropland expansion is assessments at a national or regional scale, informed by local conditions and policies (for example [104], [105]).

Implications for Policy

At the tenth Conference of the Parties to the Convention on Biological Diversity, parties agreed on 20 Aichi Biodiversity Targets [11]. These include commitments to halve, or where feasible halt, loss of natural habitats by 2020 (target 5), to ensure that areas under agriculture are managed sustainably (target 7), and to eliminate harmful pollution (target 8). The first of these can only be achieved by addressing the drivers of habitat loss, which in many parts of the world include crop expansion. To avoid compromising the first commitment, the remaining two will require yields to increase in parts of the world where productivity is currently low and where potential exists to do this without negative environmental impacts [106]. In all cases it will be difficult to reduce threats to biodiversity without strengthening public policy, such as national-level land use policies for stabilising the agricultural frontier around the last big blocks of wilderness through appropriate strategic land-use planning, infrastructure planning, better regulation of large international land acquisitions, and protected area designation.

In addition to government-led policies and incentives, there is potential for voluntary certification and other market-based initiatives to help reduce the impact of agriculture. However, this potential has been realised to only a limited extent to date. For some of the major tropical crops, including oil palm, soybeans and sugar cane, commodity roundtables have been set up to decide and implement standards for environmentally and socially responsible production [107]. These initiatives typically involve representatives from throughout the commodity chain (lenders, growers, manufacturers and retailers) as well as from governments and civil society. There is increasing attention towards the complexities and challenges of tailoring such standards to cater for small-scale farmers [108], [109]. However, these voluntary initiatives do not apply to all producers within a single country or commodity chain, and they do not have the power to implement land-use planning on the scale needed to prevent incursions of agriculture into large blocks of natural habitat, such as those in the Congo Basin. It is difficult to see how that could be achieved without government intervention, backed up by technical and financial support from wealthier governments for which tropical countries are an important source of imports and of climate-regulating ecosystem services.

All of the targets mentioned will become easier to achieve if global consumption of agricultural products can be reduced or stabilised. In the developed world, there is considerable scope to eliminate over-consumption, promote diets which are less land-demanding and reduce post-consumer waste [110], [111]. Reforming incentives for bioenergy to support only those feedstocks not implicated in direct or indirect land-use change could help to reduce global demand for agricultural land [112], [113]. In the developing world, the most important issues include rising meat consumption by an emerging middle class, rapid population growth and post-harvest losses [110], [111], [114]. Some of these issues can only be addressed by national and international policy, while others can be addressed at a local level, for example by NGOs.

Irrespective of such measures to limit over-consumption and wastage, strengthened efforts to protect wild lands from conversion will be essential if the threat of agricultural expansion to tropical biodiversity is to be reduced. Other studies have suggested that habitat conversion, once initiated in an area, is contagious and difficult to stop [115], and also that conservation in remote, less-developed parts of the world is often very cost-effective compared to conservation in more-developed areas [116]. Increasing conservation efforts in Frontier Forests and High Biodiversity Wilderness Areas may thus merit greater attention from conservationists and policy-makers.

Overlap between Cultivation Potential and Protected Areas for (A) Neotropical Countries, (B) Tropical Africa and (C) Tropical Asia/Australia

Map of cultivation potential in relation to cropland is as for Figure 5. Protected areas comprise protected areas of all types with polygon information, extracted from the 2010 version of the World Database on Protected Areas (WDPA). [The WDPA is a joint product of IUCN and UNEP prepared by UNEP-WCMC and the IUCN-WCPA working with Governments, the Secretariats of Multilateral Environmental Agreements, collaborating Non-Government Organizations and individuals.

ACKNOWLEDGMENTS

We thank Ariel Brunner, Abi Bunker, Ellie Crane and Heather Ducharme for insightful comments, and the participants of a workshop held in January 2010.

AUTHOR CONTRIBUTIONS

Analyzed the data: BP. Wrote the paper: BP MB SHMB PFD JPWS AJS AB.

REFERENCES

1. Tilman D, Fargione J, Wolff B, D'Antonio C, Dobson A, et al. (2001) Forecasting agriculturally driven global environmental change. Science 292: 281–284. doi: 10.1126/science.1057544
2. Foley JA, DeFries R, Asner GP, Barford C, Bonan G, et al. (2005) Global consequences of land use. Science 309: 570–574. doi: 10.1126/science.1111772
3. Ellis EC, Ramankutty N (2008) Putting people in the map: anthropogenic biomes of the world. Front Ecol Environ 6: 439–447. doi: 10.1890/070062
4. Haberl H, Erb KH, Krausmann F, Gaube V, Bondeau A, et al. (2007) Quantifying and mapping the human appropriation of net primary production in earth's terrestrial ecosystems. Proc Natl Acad Sci USA 104: 12942–12947. doi: 10.1073/pnas.0704243104
5. Ramankutty N, Evan AT, Monfreda C, Foley JA (2008) Farming the planet: 1. Geographic distribution of global agricultural lands in the year 2000. Global Biogeochem Cycles 22: GB1003 doi:10.1029/2007GB002952.
6. Wright HL, Lake IR, Dolman PM (2012) Agriculture – a key element for conservation in the developing world. Conserv Lett 5: 11–19 doi:10.1111/j.1755-263X.2011.00208.x.

7. Sala OE, Chapin FS, Armesto JJ, Berlow E, Bloomfield J, et al. (2000) Global biodiversity scenarios for the year 2100. Science 287: 1770–1774. doi: 10.1126/science.287.5459.1770
8. Green RE, Cornell SJ, Scharlemann JPW, Balmford A (2005) Farming and the fate of wild nature. Science 307: 550–555 doi:10.1126/science.1106049.
9. Hoffmann M, Hilton-Taylor C, Angulo A, Bohm M, Brooks TM, et al. (2010) The impact of conservation on the status of the world's vertebrates. Science 330: 1503–1509 doi:10.1126/science.1194442.
10. Donald PF (2004) Biodiversity impacts of some agricultural commodity production systems. Conserv Biol 18: 17–37.
11. UNEP (2011) Report of the Tenth Meeting of the Conference of the Parties to the Convention on Biological Diversity. Montreal, Canada: UN Environment Programme, UNEP/CBD/COP/10/27. 353 p.
12. Clay J (2004) World Agriculture and the Environment: A Commodity-by-Commodity Guide to Impacts and Practices. Washington, DC: Island Press. 570 p.
13. Scharlemann JPW, Green RE, Balmford A (2004) Land-use trends in Endemic Bird Areas: global expansion of agriculture in areas of high conservation value. Glob Change Biol 10: 2046–2051. doi: 10.1111/j.1365-2486.2004.00860.x
14. FAO (2002) World Agriculture: towards 2015/2030. 97 p.
15. OECD/FAO (2012) OECD-FAO Agricultural Outlook 2012–2021, OECD Publishing. 283 p. Available: http://dx.doi.org/10.1787/agr_outlook-2012-en. Accessed 2012 Jul 19.
16. McIntyre BD, Herren HR, Wakhungu J, Watson RT (2009) International Assessment of Agricultural Knowledge, Science and Technology for Development (IAASTD): global report, Washington DC: Island Press. 593 p. Available:http://www.agassessment.org/docs/IAASTD_EXEC_SUMMARY_JAN_2008.pdf. Accessed 2010 Mar 29.
17. Gibbs HK, Ruesch AS, Achard F, Clayton MK, Holmgren P, et al. (2010) Tropical forests were the primary sources of new agricultural land in the 1980s and 1990s. Proc Natl Acad Sci USA 107: 16732–16737 doi:10.1073/pnas.0910275107.
18. de Araujo LM, de Barros Prado Moura F (2011) Bioethanol's dirty footprint in Brazil. Nature 469: 299 doi:10.1038/469299d.
19. Koh LP (2007) Potential habitat and biodiversity losses from intensified biodiesel feedstock production. Conserv Biol 21: 1373–1375. doi: 10.1111/j.1523-1739.2007.00771.x

20. Melillo JM, Reilly JM, Kicklighter DW, Gurgel AC, Cronin TW, et al. (2009) Indirect emissions from biofuels: how important? Science 326: 1397–1399 doi:10.1126/science.1180251.

21. Miles L, Kapos V (2008) Reducing greenhouse gas emissions from deforestation and forest degradation: global land-use implications. Science 320: 1454–1455. doi: 10.1126/science.1155358

22. Geist HJ, Lambin EF (2002) Proximate causes and underlying driving forces of tropical deforestation. Bio Science 52: 143–150. doi: 10.1641/0006-3568(2002)052[0143:pcaudf]2.0.co;2

23. Phalan B, Balmford A, Green RE, Scharlemann JPW (2011) Minimising the harm to biodiversity of producing more food globally. Food Policy 36: S62–S71 doi:10.1016/j.foodpol.2010.11.008.

24. Phalan B, Onial M, Balmford A, Green RE (2011) Reconciling food production and biodiversity conservation: land sharing and land sparing compared. Science 333: 1289–1291 doi:10.1126/science.1208742.

25. Chomitz KM (2006) At Loggerheads? Agricultural Expansion, Poverty Reduction, and Environment in the Tropical Forests. Washington DC: World Bank. 284 p.

26. Laurance WF (2007) Have we overstated the tropical biodiversity crisis? Trends Ecol Evol 22: 65–70. doi: 10.1016/j.tree.2006.09.014

27. Cohen JE (2003) Human population: the next half century. Science 302: 1172–1175 doi:10.1126/science.1088665.

28. Hansen MC, Stehman SV, Potapov PV (2010) Quantification of global gross forest cover loss. Proc Natl Acad Sci USA 107: 8650–8655 doi:10.1073/pnas.0912668107.

29. Olson DM, Dinerstein E, Wikramanayake ED, Burgess ND, Powell GVN, et al. (2001) Terrestrial ecoregions of the world: A new map of life on Earth. BioScience 51: 933–938. doi: 10.1641/0006-3568(2001)051[0933:teotwa]2.0.co;2

30. Brooks TM, Mittermeier RA, da Fonseca GAB, Gerlach J, Hoffmann M, et al. (2006) Global biodiversity conservation priorities. Science 313: 58–61. doi: 10.1126/science.1127609

31. Hoekstra JM, Boucher TM, Ricketts TH, Roberts C (2005) Confronting a biome crisis: global disparities of habitat loss and protection. Ecol Lett 8: 23–29 doi:10.1111/j.1461-0248.2004.00686.x.

32. Mittermeier RA, Gil PR, Hoffmann M, Pilgrim J, Brooks T, et al.. (2004) Hotspots Revisited. Mexico City: Cemex. 392 p.

33. Stattersfield AJ, Crosby MJ, Long AJ, Wege DC (1998) Endemic Bird Areas of the World: Priorities for Biodiversity Conservation. BirdLife

Conservation Series no. 7. Cambridge, UK: BirdLife International. 846 p.

34. WWF IUCN (1997) Centres of Plant Diversity. Gland, Switzerland: WWF and IUCN. 3 vols.
35. Mittermeier RA, Robles Gil P, Mittermeier CG (1997) Megadiversity. Mexico City: CEMEX. 501 p.
36. Olson DM, Dinerstein E (2002) The Global 200: Priority Ecoregions for global conservation. Ann Missouri Bot Gard 89: 199–224 doi:10.2307/3298564.
37. Mittermeier RA, Mittermeier CG, Brooks TM, Pilgrim JD, Konstant WR, et al. (2003) Wilderness and biodiversity conservation. Proc Natl Acad Sci USA 100: 10309–10313 doi:10.1073/pnas.1732458100.
38. Bryant D, Nielsen D, Tangley L (1997) Last frontier forests: Ecosystems and economies on the edge. Washington DC: World Resources Institute. 54 p.
39. Sanderson EW, Jaiteh M, Levy MA, Redford KH, Wannebo AV, et al. (2002) The Human Footprint and the Last of the Wild. Bio Science 52: 891–904 doi:10.1641/0006-3568(2002)052[0891:THFATL]2.0.CO;2.
40. Kaptué Tchuenté AT, Roujean J-L, De Jong SM (2011) Comparison and relative quality assessment of the GLC2000, GLOBCOVER, MODIS and ECOCLIMAP land cover data sets at the African continental scale. Int J Appl Earth Obs 13: 207–219 doi:16/j.jag.2010.11.005.
41. Herold M, Mayaux P, Woodcock CE, Baccini A, Schmullius C (2008) Some challenges in global land cover mapping: An assessment of agreement and accuracy in existing 1 km datasets. Remote Sens Environ 112: 2538–2556 doi:10.1016/j.rse.2007.11.013.
42. Fritz S, See L (2008) Identifying and quantifying uncertainty and spatial disagreement in the comparison of Global Land Cover for different applications. Glob Change Biol 14: 1057–1075. doi: 10.1016/j.rse.2007.11.013
43. McCallum I, Obersteiner M, Nilsson S, Shvidenko A (2006) A spatial comparison of four satellite derived 1 km global land cover datasets. Int J Appl Earth Obs 8: 246–255 doi:10.1016/j.jag.2005.12.002.
44. Monfreda C, Ramankutty N, Foley JA (2008) Farming the planet: 2. Geographic distribution of crop areas, yields, physiological types, and net primary production in the year 2000. Global Biogeochem Cycles 22: GB1022. doi: 10.1016/j.rse.2007.11.013
45. You L, Guo Z, Koo J, Ojo W, Sebastian K, et al.. (2010) Spatial Production Allocation Model (SPAM) 2000 Version 3 Release 2. HarvestChoice. Available:http://MapSPAM.info. Accessed: 2010 Sep 27.

46. FAOSTAT Statistical databases. Food and Agriculture Organization of the United Nations. Available: http://faostat.fao.org/. Accessed: 2011 Jan 11.

47. Ewers RM, Scharlemann JPW, Balmford A, Green RE (2009) Do increases in agricultural yield spare land for nature? Global Change Biol 15: 1716–1726 doi:10.1111/j.1365-2486.2009.01849.x.

48. Thoenes P (2011) FAO price & policy bulletin, January 2011. Rome, Italy: Food and Agriculture Organization of the United Nations. 4 p.

49. Dávalos LM, Bejarano AC, Hall MA, Correa HL, Corthals A, et al. (2011) Forests and drugs: coca-driven deforestation in tropical biodiversity hotspots. Environ Sci Technol 45: 1219–1227 doi:10.1021/es102373d. View Article PubMed/NCBI Google Scholar

50. Ramankutty N (2004) Croplands in West Africa: A geographically explicit dataset for use in models. Earth Interact 8: 1–22 doi:10.1175/1087-3562(2004)8<1:CIWAAG>2.0.CO;2.

51. FAO (2011) Crops statistics. Concepts, definitions and classifications. Food and Agriculture Organization of the United Nations. Available:http://www.fao.org/fileadmin/templates/ess/ess_test_folder/documents/Production_trade/definitions/Crops_statistics_concepts_definitions_classifications.doc. Accessed: 2011 May 31.

52. Cassman KG (1999) Ecological intensification of cereal production systems: Yield potential, soil quality, and precision agriculture. Proc Natl Acad Sci USA 96: 5952–5959. doi: 10.1073/pnas.96.11.5952

53. FAO (2010) World Programme for the Census of Agriculture. Part two: agricultural census items – concepts, definitions and tabulations. Rome, Italy: Food and Agriculture Organization of the United Nations. 121 p. Available:http://www.fao.org/fileadmin/templates/ess/documents/world_census_of_agriculture/appendix3_r7.pdf. Accessed: 2011 Apr 11.

54. FAO (2011) Tree crops. Guidelines for estimating area data. Rome, Italy: Food and Agriculture Organization of the United Nations. Available:http://www.fao.org/fileadmin/templates/ess/ess_test_folder/documents/Production_trade/definitions/Tree_crops_guidelines_for_estaimating_area.doc. Accessed: 2011 Jul 29.

55. Tóth G, Kozlowski B, Prieler S, Wiberg D (2011) Global Agro-Ecological Zones (GAEZ v3.0) User's Guide. Laxenburg and Rome: IIASA and FAO. 56 p. Available:http://www.iiasa.ac.at/Research/LUC/GAEZv3.0/docs/GAEZ_User_Guide.pdf. Accessed: 2011 Jun 28.

56. Tabor K, Williams JW (2010) Globally downscaled climate projections for assessing the conservation impacts of climate change. Ecol Appl 20: 554–565. doi: 10.1890/09-0173.1

57. Fearnside PM (2001) Soybean cultivation as a threat to the environment in Brazil. Environmental Conservation 28: 23–38 doi:10.1017/S0376892901000030.
58. Spehar CR (1995) Impact of strategic genes in soybean on agricultural development in the Brazilian tropical savannahs. Field Crop Res 41: 141–146 doi:16/0378-4290(95)00007-D.
59. Aldhous P (2004) Borneo is burning. Nature 432: 144–146. doi: 10.1038/432144a
60. IMAGE Team (2001) The IMAGE 2.2 implementation of the SRES Scenarios. A comprehensive analysis of emissions, climate change and impacts in the 21st century. Bilthoven: Netherlands Environmental Assessment Agency (MNP) CD-ROM Publication 500110001.
61. Dobrovolski R, Diniz-Filho JAF, Loyola RD, Marco Júnior P (2011) Agricultural expansion and the fate of global conservation priorities. Biodivers Conserv 20: 2445–2459 doi:10.1007/s10531-011-9997-z.
62. Hansen MC, Stehman SV, Potapov PV, Loveland TR, Townshend JRG, et al. (2008) Humid tropical forest clearing from 2000 to 2005 quantified by using multitemporal and multiresolution remotely sensed data. Proc Natl Acad Sci USA 105: 9439–9444. doi: 10.1073/pnas.0804042105
63. Barona E, Ramankutty N, Hyman G, Coomes OT (2010) The role of pasture and soybean in deforestation of the Brazilian Amazon. Environ Res Lett 5: 024002 doi:10.1088/1748-9326/5/2/024002.
64. FAO (2010) World Programme for the Census of Agriculture. Appendix 3: classification of crops. Rome, Italy: Food and Agriculture Organization of the United Nations. 5 p. Available: http://www.fao.org/fileadmin/templates/ess/documents/world_census_of_agriculture/appendix3_r7.pdf. Accessed: 2011 Apr 11.
65. Fischer G, van Velthuizen H, Shah M, Nachtergaele F (2002) Global Agro-ecological Assessment for Agriculture in the 21st Century: Methodology and Results. Laxenburg, Austria and Rome, Italy: International Institute for Applied Systems Analysis/Food and Agriculture Organization of the United Nations. 119 p.
66. ESRI 2009. ArcGIS Desktop: Release 9.3. Redlands, CA: Environmental Systems Research Institute.
67. Baudron F, Corbeels M, Andersson JA, Sibanda M, Giller KE (2011) Delineating the drivers of waning wildlife habitat: The predominance of cotton farming on the fringe of protected areas in the Mid-Zambezi Valley, Zimbabwe. Biol Conserv 144: 1481–1493 doi:10.1016/j.biocon.2011.01.017.

68. Rappole JH, King DI, Rivera JHV (2003) Coffee and conservation. Conserv Biol 17: 334–336 doi:10.1046/j.1523-1739.2003.01548.x.
69. Yang L, Pan R, Wang S (1985) Investigation on the birds of cultivated land of tea trees and rubber trees in Xishuangbanna, Yunnan province. Zool Res 4: 353–360. doi: 10.1016/j.biocon.2011.01.017
70. Schroth G, Harvey CA (2007) Biodiversity conservation in cocoa production landscapes: an overview. Biodivers Conserv 16: 2237–2244. doi: 10.1016/j.biocon.2011.01.017
71. Li H, Aide TM, Ma Y, Liu W, Cao M (2007) Demand for rubber is causing the loss of high diversity rain forest in SW China. Biodivers Conserv 16: 1731–1745.
72. Carrere R, Lohmann L (1996) Pulping the South: Industrial Tree Plantations in the World Paper Economy. London and New Jersey: Zed Books. 208 p. Available:http://www.wrm.org.uy/plantations/material/PulpingSouth.pdf. Accessed 2009 Apr 10.
73. Wilcove D, Koh L (2010) Addressing the threats to biodiversity from oil-palm agriculture. Biodivers Conserv 19: 999–1007 doi:10.1007/s10531-009-9760-x.
74. BirdLife International (2011) BirdLife species factsheets. Available:http://www.birdlife.org. Accessed 2011 Aug 4.
75. Scales IR (2011) Farming at the forest frontier: land use and landscape change in western Madagascar, 1896–2005. Environ Hist 17: 499–524 doi:10.3197/096734011X13150366551481.
76. Seddon N, Butchart S, Tobias J, Yount JW, Remi Ramanampamonjy J, et al. (2000) Conservation issues and priorities in the Mikea Forest of south-west Madagascar. Oryx 34: 287–304 doi:10.1046/j.1365-3008.2000.00134.x.
77. Balmford A, Green R, Phalan B (2012) What conservationists need to know about farming. Proc R Soc B Available: http://rspb.royalsocietypublishing.org/content/early/2012/04/24/rspb.2012.0515.
78. Butler RA, Laurance WF (2008) New strategies for conserving tropical forests. Trends Ecol Evol 23: 469–472. doi: 10.1016/j.tree.2008.05.006
79. Duveiller G, Defourny P, Desclée B, Mayaux P (2008) Deforestation in Central Africa: Estimates at regional, national and landscape levels by advanced processing of systematically-distributed Landsat extracts. Remote Sens Environ 112: 1969–1981 doi:10.1016/j.rse.2007.07.026.
80. Zhang Q, Devers D, Desch A, Justice CO, Townshend J (2005) Mapping tropical deforestation in Central Africa. Environ Monit Assess 101: 69–83. doi: 10.1046/j.1365-3008.2000.00134.x

81. FAO (2006) Global Forest Resources Assessment 2005: progress towards sustainable forest management. Rome, Italy: Food and Agriculture Organization of the United Nations. 320 p.
82. Rudel TK (2005) Tropical forests: regional paths of destruction and regeneration in the late twentieth century. New York, USA: Columbia Univ Pr. 231 p.
83. Wolfire DM, Brunner J, Sizer N (1998) Forests and the Democratic Republic of Congo. Opportunity in a Time of Crisis. Washington DC: World Resources Institute. 28 p. Available: http://www.globalforestwatch.org/common/pdf/drc_eng.pdf. Accessed: 2011 Dec 5.
84. United Nations Population Division (2011) World Population Prospects, the 2010 Revision. United Nations. Available: http://esa.un.org/unpd/wpp/unpp/panel_population.htm. Accessed: 2011 Dec 6.
85. Laporte NT, Stabach JA, Grosch R, Lin TS, Goetz SJ (2007) Expansion of industrial logging in central Africa. Science 316: 1451 doi:10.1126/science.1141057.
86. Glaser B, Haumaier L, Guggenberger G, Zech W (2001) The 'Terra Preta' phenomenon: a model for sustainable agriculture in the humid tropics. Naturwissenschaften 88: 37–41 doi:10.1007/s001140000193.
87. Sanchez PA (2002) Soil fertility and hunger in Africa. Science 295: 2019–2020 doi:10.1126/science.1065256.
88. Karsenty A (2010) Large-Scale Acquisition of Rights on Forest Lands in Africa. Washington DC: Rights and Resources Initiative. 20 p.
89. Wilkie D, Shaw E, Rotberg F, Morelli G, Auzel P (2000) Roads, development, and conservation in the Congo Basin. Conserv Biol 14: 1614–1622 doi:10.2307/2641513.
90. Wunder S (2004) Policy options for stabilising the forest frontier: a global perspective. In: Gerold G, Fremerey M, Guhardja E, editors. Land use, nature conservation and the stability of rainforest margins in Southeast Asia. Berlin, Germany: Springer-Verlag. pp. 3–25.
91. Brown HCP, Smit B, Sonwa DJ, Somorin OA, Nkem J (2011) Institutional perceptions of opportunities and challenges of REDD+ in the Congo Basin. J Environ Devel 20: 381–404 doi:10.1177/1070496511426480.
92. Nepstad D, Soares-Filho BS, Merry F, Lima A, Moutinho P, et al. (2009) The end of deforestation in the Brazilian Amazon. Science 326: 1350–1351 doi:10.1126/science.1182108.
93. Soares-Filho B, Moutinho P, Nepstad D, Anderson A, Rodrigues H, et al. (2010) Role of Brazilian Amazon protected areas in climate change mitigation. Proc Natl Acad Sci USA 107: 10821–10826 doi:10.1073/pnas.0913048107.

94. Martinelli LA (2011) Block changes to Brazil's Forest Code. Nature 474: 579 doi:10.1038/474579a.

95. Nepstad D, Schwartzman S, Bamberger B, Santilli M, Ray D, et al. (2006) Inhibition of Amazon deforestation and fire by parks and indigenous lands. Conserv Biol 20: 65–73. doi: 10.1111/j.1523-1739.2006.00351.x

96. Cook G (2009) Chapter 6. Historical perspectives on land use development in northern Australia: with emphasis on the Northern Territory. Northern Australia Land and Water Science Review full report. Available: http://www.nalwt.gov.au/files/Chapter_06-Historical_perspectives_on_land_use_development_in_north.pdf. Accessed 2012 Jan 25.

97. Deininger K, Byerlee D, Lindsay J, Norton A, Selod H, et al.. (2011) Rising global interest in farmland. Washington, DC: World Bank. Available:http://siteresources.worldbank.org/DEC/Resources/Rising-Global-Interest-in-Farmland.pdf. Accessed: 2011 Jan 7.

98. Brazilian agriculture: The miracle of the cerrado (2010) The Economist. Available:http://www.economist.com/node/16886442. Accessed 2010 Aug 29.

99. Lapola DM, Schaldach R, Alcamo J, Bondeau A, Koch J, et al. (2010) Indirect land-use changes can overcome carbon savings from biofuels in Brazil. Proc Natl Acad Sci USA 107: 3388–3393 doi:10.1073/pnas.0907318107.

100. Klink CA, Machado RB (2005) Conservation of the Brazilian Cerrado. Conserv Biol 19: 707–713 doi:10.1111/j.1523-1739.2005.00702.x.

101. Galford GL, Melillo J, Mustard JF, Cerri CE, Cerri CC (2010) The Amazon frontier of land-use change: croplands and consequences for greenhouse gas emissions. Earth Interact 14: 15. doi: 10.1175/2010ei327.1

102. NEMA (2010) The potential of bio-fuel in Uganda: an assessment of land resources for bio-fuel feedstock suitability. Uganda: National Environment Management Authority. 39 p.

103. Muok BO, Nyabenge M, Ouma BO, Esilaba AO, Nandokha T, et al.. (2010) Environmental Suitability and Agro-environmental Zoning of Kenya for Biofuel Production. Nairobi, Kenya: African Centre for Technology Studies, Policy Innovation Systems for Clean Energy Security and United Nations Environmental Programme. 107 p.

104. Mertens B, Lambin EF (2000) Land-cover-change trajectories in southern Cameroon. A Assoc Am Geog 90: 467–494 doi:10.1111/0004-5608.00205.

105. Müller R, Pistorius T, Rohde S, Gerold G, Pacheco P (2013) Policy options to reduce deforestation based on a systematic analysis of drivers and agents in lowland Bolivia. Land Use Policy 30: 895–907 doi:10.1016/j.landusepol.2012.06.019.

106. Tilman D, Balzer C, Hill J, Befort BL (2011) Global food demand and the sustainable intensification of agriculture. Proc Natl Acad Sci USA 108: 20260–20264 doi:10.1073/pnas.1116437108.

107. Scarlat N, Dallemand J-F (2011) Recent developments of biofuels/ bioenergy sustainability certification: A global overview. Energy Policy 39: 1630–1646 doi:10.1016/j.enpol.2010.12.039.

108. Gómez MI, Barrett CB, Buck LE, De Groote H, Ferris S, et al. (2011) Research principles for developing country food value chains. Science 332: 1154–1155 doi:10.1126/science.1202543.

109. Lee JSH, Rist L, Obidzinski K, Ghazoul J, Koh LP (2011) No farmer left behind in sustainable biofuel production. Biol Conserv 144: 2512–2516 doi:16/j.biocon.2011.07.006.

110. Parfitt J, Barthel M, Macnaughton S (2010) Food waste within food supply chains: quantification and potential for change to 2050. Philos T Roy Soc B 365: 3065–3081 doi:10.1098/rstb.2010.0126.

111. Wirsenius S, Azar C, Berndes G (2010) How much land is needed for global food production under scenarios of dietary changes and livestock productivity increases in 2030? Agr Syst 103: 621–638 doi:16/j.agsy.2010.07.005.

112. Hellmann F, Verburg PH (2010) Impact assessment of the European biofuel directive on land use and biodiversity. J Environ Man 91: 1389–1396 doi:10.1016/j.jenvman.2010.02.022.

113. Havlík P, Schneider UA, Schmid E, Böttcher H, Fritz S, et al. (2011) Global land-use implications of first and second generation biofuel targets. Energy Policy 39: 5690–5702. doi: 10.1016/j.agsy.2010.07.005

114. Bongaarts J, Sinding S (2011) Population policy in transition in the developing world. Science 333: 574–576 doi:10.1126/science.1207558.

115. Boakes EH, Mace GM, McGowan PJK, Fuller RA (2010) Extreme contagion in global habitat clearance. Proc Roy Soc B 277: 1081–1085. doi: 10.1016/j.agsy.2010.07.005

116. Balmford A, Gaston KJ, Blyth S, James A, Kapos V (2003) Global variation in terrestrial conservation costs, conservation benefits, and unmet conservation needs. Proc Natl Acad Sci USA 100: 1046–1050. doi: 10.1073/pnas.0236945100

Chapter 3

INCREASING CROPPING SYSTEM DIVERSITY BALANCES PRODUCTIVITY, PROFITABILITY, AND ENVIRONMENTAL HEALTH

[1]Adam S. Davis mail, [2]Jason D. Hill, [3]Craig A. Chase, [4]Ann M. Johanns, [5]Matt Liebman

[1]United States Department of Agriculture/Agricultural Research Service, Global Change and Photosynthesis Research Unit, Urbana, Illinois, United States of America.

[2]Department of Bioproducts and Biosystems Engineering, University of Minnesota, St. Paul, Minnesota, United States of America.

[3]Leopold Center for Sustainable Agriculture, Iowa State University, Ames, Iowa, United States of America.

[4]Department of Economics, Iowa State University Extension and Outreach, Osage, Iowa, United States of America.

[5]Department of Agronomy, Iowa State University, Ames, Iowa, United States of America.

ABSTRACT

Balancing productivity, profitability, and environmental health is a key challenge for agricultural sustainability. Most crop production systems in the United States are characterized by low species and management diversity, high use of fossil energy and agrichemicals, and large negative impacts on the environment. We hypothesized that cropping system diversification would promote ecosystem services that would supplement, and eventually displace, synthetic external inputs used to maintain crop productivity. To test this, we conducted a field study from 2003–2011 in Iowa that included three contrasting systems varying in length of crop sequence and inputs. We compared a conventionally managed 2-yr rotation (maize-soybean) that received fertilizers and herbicides at rates comparable to those used on nearby farms with two more diverse cropping systems: a 3-yr rotation (maize-soybean-small grain + red clover) and a 4-yr rotation (maize-soybean-small grain + alfalfa-alfalfa) managed with lower synthetic N fertilizer and herbicide inputs and periodic applications of cattle manure. Grain yields, mass of harvested products, and profit in the more diverse systems were similar to, or greater than, those in the conventional system, despite reductions of agrichemical inputs. Weeds were suppressed effectively in all systems, but freshwater toxicity of the more diverse systems was two orders of magnitude lower than in the conventional system. Results of our study indicate that more diverse cropping systems can use small amounts of synthetic agrichemical inputs as powerful tools with which to tune, rather than drive, agroecosystem performance, while meeting or exceeding the performance of less diverse systems.

INTRODUCTION

One of the key challenges of the 21st century is developing ways of producing sufficient amounts of food while protecting both environmental quality and the economic well-being of rural communities [1], [2]. Over the last half century, conventional approaches to crop production have relied heavily on manufactured fertilizers and pesticides to increase yields, but they have also degraded water quality and posed threats to human health and wildlife [3]–[6]. Consequently, attention is now being directed toward the development of crop production systems with improved resource use efficiencies and more benign effects on the environment [1], [7]. Less attention has been paid to developing better methods of pest management, especially for weeds. Here we explore the potential benefits of diversifying cropping systems as a means of controlling weed population dynamics while simultaneously enhancing other desirable agroecosystem processes [8]. We focus on crop rotation, an approach to cropping system diversification whereby different species are placed in the same field at different times.

Rotation systems have been used for millennia to maintain soil fertility and productivity and to suppress pests, and can increase yields even in situations where substantial amounts of fertilizers and pesticides are applied [9], [10]. Rotation systems also foster spatial diversity, since different crops within the rotation sequence are typically grown in different fields on a farm in the same year. Diversification through crop rotation can be an especially useful strategy in farming systems that integrate crop and livestock production. The addition of forage crops, including turnips and clovers, to cereal-based systems in northwestern Europe and England in the 1600s and 1700s enhanced nitrogen supply through fixation by legumes, and increased nutrient cycling due to greater livestock density and manure production. These changes allowed the intensification of both crop and livestock production and increased yields substantially [11], [12]. Integrated crop–livestock systems remained widespread in northern Europe, England, and much of the humid, temperate regions of North America until the 1950s and 1960s, when increased availability of relatively low-cost synthetic fertilizers made mixed farming and nutrient recycling biologically unnecessary and specialized crop and livestock production more economically attractive. In recent years, there has been interest in reintegrating crop and livestock systems as a strategy for reducing reliance on fossil fuels, minimizing the use of increasingly expensive fertilizers, and limiting water pollution by nutrients, pathogens, and antibiotics [13], [14].

Weeds are a ubiquitous and recurrent problem in essentially all crop production systems, and chemicals applied for weed control dominate the world market for pesticides [15]. With the introduction of crop genotypes engineered to tolerate herbicides, especially glyphosate, and with the continuing availability of older, relatively inexpensive herbicides, such as atrazine, successful weed management in conventional crop production systems has been largely taken for granted since the mid-1990s. Now, however, with expanded recognition of herbicides as environmental contaminants [4] and the increasing prevalence of herbicide resistant weeds [16], there is an important need to develop weed management strategies that are less reliant on herbicides and that subject weeds to a wide range of stress and mortality factors [17]. We believe that cropping system diversification may play an important role in the development of such strategies.

Here, we report the results of a large-scale, long-term experiment examining the consequences of cropping system diversification on agronomic, economic, and environmental measures of system performance. The experiment was conducted during 2003–2011 in Boone County, Iowa, within the central U.S. maize production region, and comprised three contrasting cropping systems varying in length of crop sequence, levels of chemical inputs, and use of manure. We compared a conventionally managed 2-yr rotation (maize-soybean) that received fertilizers and herbicides at rates comparable to those used on surrounding commercial farms with two more diverse cropping systems: a 3-yr rotation (maize-soybean-small grain + red clover) and a 4-yr rotation (maize-soybean-small grain + alfalfa-alfalfa) managed with reduced N fertilizer and herbicide inputs and periodic applications of composted cattle manure. Triticale was used as the small grain crop in 2003–2005; oat was used in 2006–2011. The 2-yr rotation is typical of cash grain farming systems in the region, whereas the 3-yr and 4-yr rotations are representative of farming systems in the region that include livestock. Details of the experimental site, management practices, sampling procedures, and data analyses are provided in the online SI section (Text S1, Figure S1, Tables S1-S4).

A central hypothesis framing our study was that cropping system diversification would result in the development of ecosystem services over time that would supplement, or eventually displace, the role of synthetic external inputs in maintaining crop productivity and profitability. Based on this hypothesis, we predicted that input requirements of the more diverse systems would initially be similar to that of the less diverse system, but would increasingly diverge from the less diverse system over time as the systems matured. We also predicted that crop yields, weed suppression, and economic performance of the three systems would be similar throughout the study. Finally, we predicted that reduced requirements for external synthetic inputs for pest management would result in a lower toxicological profile of the more diverse systems compared to the less diverse system.

RESULTS

Crop Yields and Net Profitability

Cropping system diversification enhanced yields of maize and soybean grain and system-level harvested crop mass (grain, straw, and hay) while maintaining economic returns. The most parsimonious linear statistical models for each of these measures of system performance contained terms for main effects of year and system, but no interaction term (AICwith interaction = 319; AIC no interaction = 315). Over the 2003 to 2011 period, maize grain yield was on average 4% greater in the 3-yr and 4-yr rotations than in the 2-yr rotation (means for the 2-yr, 3-yr and 4-yr rotations are hereafter referred to as µ2, µ3 and µ4, respectively; µ2 = 12.3±0.1 Mg ha−1; µ3 = 12.7±0.2 Mg ha−1; µ4 = 12.9±0.2 Mg ha−1; pre-planned 1 d.f. contrast of system: F1,7 = 8, P = 0.03), and similar in the 3-yr and 4-yr rotations (Fig. 1a). Soybean grain yield during the same period was on average 9% greater in the 3-yr and 4-yr rotations than in the 2-yr rotation (µ2 = 3.4±0.07 Mg ha−1; µ3 = 3.8±0.08 Mg ha−1; µ4 = 3.8±0.08 Mg ha−1; F1,7 = 11.3, P = 0.01) and similar in the 3-yr and 4-yr rotations (Fig. 1b). Harvested crop mass, averaged over the various crop phases comprising each cropping system, followed a similar pattern to maize and soybean grain yields. Mean crop biomass for 2003 to 2011 was 8% greater in the 3-yr and 4-yr rotations than in the 2-yr rotation (µ2 = 7.9±0.08 Mg ha−1; µ3 = 8.5±0.1 Mg ha−1; µ4 = 8.6±0.2 Mg ha−1; system: t6 = 5.1, P = 0.002), and similar in the 3-yr and 4-yr rotations (Fig. 1c).

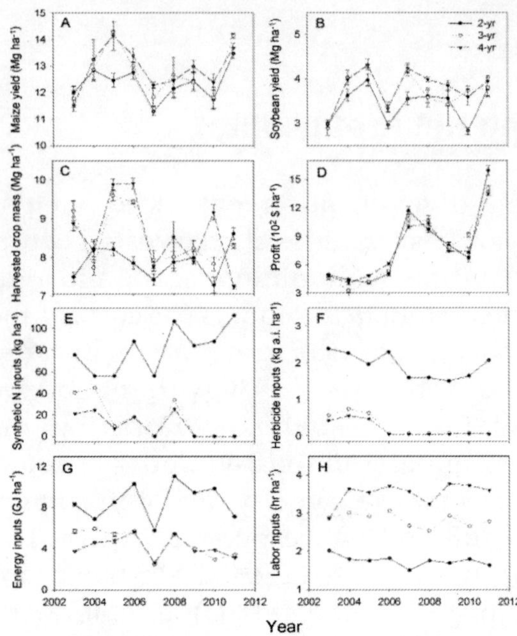

Figure 1 Cropping system performance over time. Annual performance of maize-soybean (2-yr), maize-soybean-small grain/red clover (3-yr), and maize-soybean-small grain/alfalfa-alfalfa (4-yr) cropping systems in Boone, IA, from 2003 to 2011. Performance metrics included: a) maize yield, b) soybean yield, c) rotation-level harvested crop mass, d) net returns to land and management, e) manufactured N fertilizer application rate, f) herbicide application rate, g) fossil energy use, and h) labor requirements. Symbols represent the mean ± SEM of four replicate experimental blocks (N = 36 per cropping system).

doi:10.1371/journal.pone.0047149.g001

We examined system profitability by calculating net returns to land and management, which represent profits to a farm operation without accounting for costs of land (e.g., rent or mortgage payments), management time (e.g., marketing), and federal subsidies. Profitability was analyzed for two temporal periods. From 2003 to 2005, considered the "startup" phase for the study, there were no differences among cropping systems in net profit, either through an analysis of main effects of system (μ_2 = \$448±17 ha^{-1}; μ_3 = \$402±17 ha^{-1}; μ_4 = \$457±15 ha^{-1}; $F_{2,6}$ = 0.12, P = 0.89) or by pre-planned 1-d.f. contrasts (2-yr vs. 3-yr and 4-yr rotations: $F_{1,7}$ = 0.10, P = 0.77) (Fig. 1d). From 2006 to 2011, the "established" phase of the study, there were again no differences among systems, either through main effects of system (μ_2 = \$953±36 ha^{-1}; μ_3 = \$965±34 ha^{-1}; μ_4 = \$913±26 ha^{-1}; $F_{2,6}$ = 0.62, P = 0.57) or by pre-planned 1-d.f. contrasts (2-yr vs. 3-yr and 4-yr rotations: $F_{1,7}$ = 0.03, P = 0.86).

Stability of system performance over time, as measured through a comparison of variances for the various products of the system, was similar for maize grain yield ($F_{2,6}$ = 2.4, P = 0.17), soybean grain yield ($F_{2,6}$ = 0.95, P = 0.44) and net returns to land and management during the startup phase of the study, 2003 to 2005 ($F_{2,6}$ = 0.05, P = 0.95). Two system products, harvested crop mass from 2003 to 2011 and profit during the established phase of the study, 2006 to 2011, showed considerable differences in system stability over time, but in contrasting ways. Variance in mean harvested crop mass was greater in the 3-yr and 4-yr rotations than in the 2-yr rotation (σ_2^2 = 0.27; σ_3^2 = 0.60; σ_4^2 = 0.95; $F_{1,7}$ = 16, P = 0.005). Conversely, cropping system diversification was associated with lower variance in profit during the established phase of the study. Variance in profit from 2006 to 2011 was lower in the 3-yr and 4-yr rotations than in the 2-yr rotation (σ_2^2 = 1.5×10^5; σ_3^2 = 8.1×10^3; σ_4^2 = 6.3×10^3; $F_{1,7}$ = 16, P = 0.005).

Agrichemical, Labor, and Energy Inputs

Application rates of the primary agrichemicals used in this study, manufactured N fertilizer ($F_{2,14}$ = 117, $P<0.0001$) and herbicides ($F_{2,14}$ = 287, $P<0.0001$), both showed strong effects of cropping system. Manufactured N fertilizer applications were higher in the 2-yr rotation than in the 3-yr and 4-yr rotations (μ_2 = 80±3 kg N ha^{-1}; μ_3 = 16±3 kg N ha^{-1}; μ_4 = 11±2 kg N ha^{-1}; $F_{1,17}$ = 16, P = 0.005), with the difference between systems increasing over the course of the study ($F_{2,14}$ = 11.6, P = 0.001) (Fig. 1e). Herbicide application rates followed a similar pattern, with greater amounts of herbicide applied in the 2-yr rotation than in the 3-yr and 4-yr rotations (μ_2 = 1.9±0.06 kg a.i. ha^{-1}; μ_3 = 0.26±0.05 kg a.i. ha^{-1}; μ_4 = 0.20±0.03 kg a.i. ha^{-1}; $F_{1,17}$ = 610, $P<0.0001$); differences among systems, however, did not increase over time (Fig. 1f).

Fossil energy use was strongly influenced by cropping system in both the startup ($F_{2,6}$ = 94, $P<0.0001$) and established ($F_{2,6}$ = 116, $P<0.0001$) phases of the study, with no difference in energy use between experimental phases ($F_{1,92}$ = 0.39, P = 0.53) (Fig. 1g). From 2003 to 2011, inputs of energy were greater in the 2-yr rotation than in the 3-yr and 4-yr rotations (μ_2 = 8.6±0.1 GJ ha^{-1}; μ_3 = 4.5±0.1 GJ ha^{-1}; μ_4 = 4.2±0.04 GJ ha^{-1}; $F_{1,7}$ = 55, P = 0.0001). The partial correlations between energy use in a given cropping system and energy use in the maize phase of that rotation, taking into account the amount of N fertilizer applied to maize, were 0.94, 0.81 and 0.70 in the 2-yr, 3-yr and 4-yr systems, respectively (SI, Table S5). This indicated that synthetic N fertilizer use in the maize phase of the various cropping systems drove energy use within the maize phase, which in turn drove energy use by a given cropping system.

Demand for labor differed among the three cropping systems in both the startup ($F_{2,4}$ = 26, P = 0.005) and established ($F_{2,10}$ = 299, $P<0.0001$) study phases, but followed a contrasting pattern to energy requirements (Fig. 1h). Labor inputs were more than 33% lower in the 2-yr rotation than in the 3-yr and 4-yr rotations from 2003 to 2005 ($F_{1,5}$ = 35, P = 0.002) and from 2006 to 2011 ($F_{1,11}$ = 59, $P<0.0001$). Overall, there was a strong negative correlation (r = −0.79, $P<0.0001$) between fossil energy and labor inputs over time in the three cropping systems.

Divergent Weed Management Systems

Two lines of evidence indicate that weeds were managed effectively in all three cropping systems in both the 'startup' and 'established' phases, in spite of reducing herbicide use by 88% in the 3-yr and 4-yr rotations compared to the 2-yr rotation. First, weed seedbanks declined at an equal rate in all study systems (Fig. 2a). Selection among linear mixed effects regression models incorporating temporal autocorrelation among seedbank measurements over time supported different intercepts (system: $F_{2,6}$ = 16.8, P = 0.0035) but did not support inclusion of a year by system interaction term (AIC_s = 182; AIC_{s*y} = 185), thus indicating a common slope (b_1 = −0.18). For all three systems, the time to decline to 95% of the weed seedbank levels in 2003 was 16.6 years. Declines in weed seedbanks reflected a focus of management attention on the timing of weed management activities and herbicide choices in all three systems, as well as the increased number and diversity of stress and mortality factors present in the 3-yr and 4-yr rotations [8], [21]. Higher densities of weed seeds in the 3-yr and 4-yr rotations, as indicated by their greater intercept values than for the 2-yr rotation (Fig. 2a.), were the result of poorer weed control in the 3-yr and 4-yr rotations during the set-up of the experiment plots in 2002.

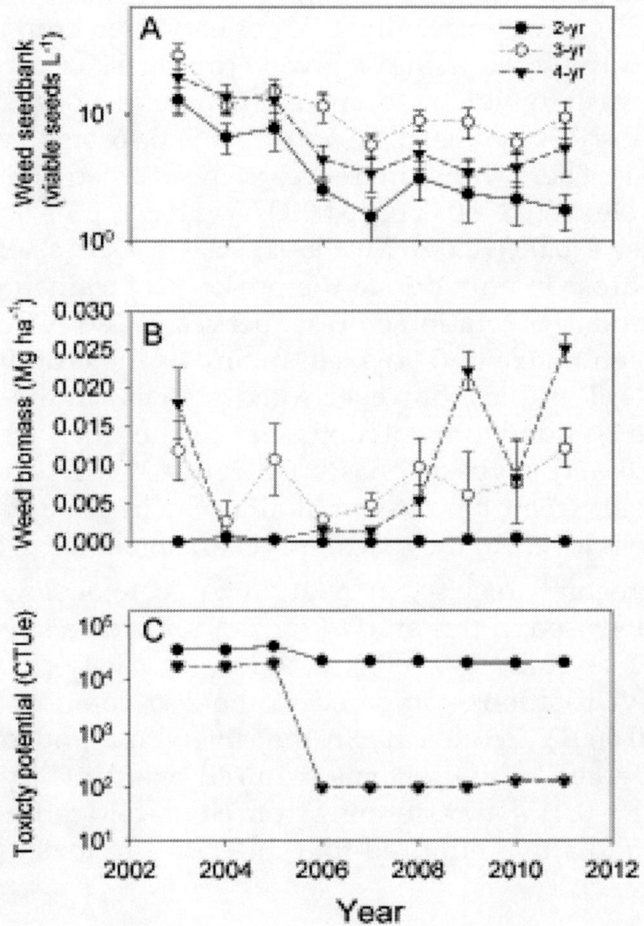

Figure 2 Divergent weed management systems. Weed management characteristics in maize-soybean (2-yr), maize-soybean-small grain/red clover (3-yr), and maize-soybean-small grain/alfalfa-alfalfa (4-yr) cropping systems in Boone, IA, from 2003 to 2011. Performance met-

rics included a) weed seed density in soil, b) weed aboveground biomass, and c) freshwater toxicity potential expressed in comparative toxic units (CTUe). Symbols represent the mean ± SEM of four replicate experimental blocks (N = 36 per cropping system).

doi:10.1371/journal.pone.0047149.g002

The second line of evidence concerns weed biomass, which was very low in all three cropping systems for the duration of the study (Fig. 2b), never exceeding 0.3% of harvested crop mass. Weed biomass was the same within a given crop phase, regardless of the cropping system in which it occurred (main effect of system: maize, $F_{2,6} = 1.47$; $P = 0.30$; soybean, $F_{2,6} = 0.88$; $P = 0.46$; small grain, $F_{1,3} = 1.24$; $P = 0.31$). There were differences in mean weed biomass among cropping systems ($\mu_2 = 0.0003 \pm 0.00007$ Mg ha^{-1}; $\mu_3 = 0.0076 \pm 0.0012$ Mg ha^{-1}; $\mu_4 = 0.009 \pm 0.001$ Mg ha^{-1}; $F_{2,6} = 12.7$; $P < 0.007$). These differences arose mainly due to the presence of a small grain phase in the 3-yr and 4-yr rotation crop sequences. Weed biomass did not differ between maize and soybean in any of the cropping systems ($F_{1,202} = 2.1$; $P = 0.15$), however weed biomass in the small grain phase of the 3-yr and 4-yr rotations was greater than weed biomass in the maize and soybean phases ($F_{1,206} = 174$; $P < 0.0001$). In the 4-year system, weed biomass in alfalfa was intermediate between weed biomass levels in the maize/soybean and small grain phases.

Environmental toxicity, in relation to ecotoxicological profiles for herbicides used in this study (Fig. 2c), showed a strong effect of system ($F_{2,14} = 1673$, $P < 0.0001$), with lower toxicity potential in the 3-yr and 4-yr rotations compared to the 2-yr rotation (type: $F_{1,17} = 2691$, $P < 0.0001$). Ecotoxicity in the diversified and conventional systems diverged as the systems matured over time [type x phase: $F_{1,16} = 7.4$, $P = 0.015$], transitioning from a two-fold difference during 2003 to 2005 to a two hundred-fold difference in toxicity from 2006 to 2011 (Fig. 2c).

DISCUSSION

Our results support the hypothesis that the development of ecosystem services over time in more diverse cropping rotations increasingly displaces the need for external synthetic inputs to maintain crop productivity. From 2003 to 2011, as predicted, the desired products (crop yield, weed suppression, and economic performance) of the more diverse and less diverse cropping rotations were similar, whereas external inputs and environmental impacts differed greatly among the systems (Fig. 3). Comparing these metrics of system performance by experimental phase (initial three years of system establishment versus the following six years) confirmed our prediction that system inputs and environmental impacts would diverge over time, whereas yield and profit would remain similar among more diverse and less diverse rotations. In the more diverse rotations, small amounts of synthetic agrichemical inputs thus served as powerful tools with which to tune, rather than drive, agroecosystem performance.

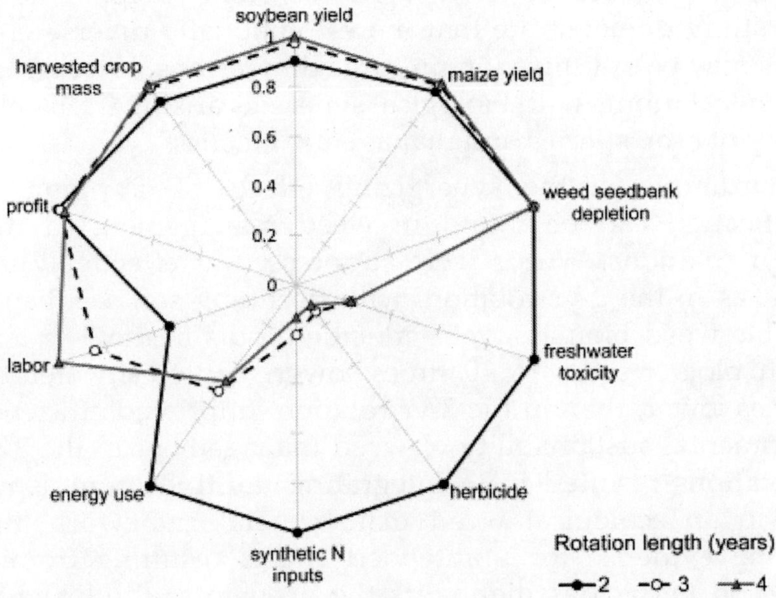

Figure 3 Multiple indicators of cropping system performance. Comparative long-term performance of maize-soybean (2-yr), maize-soybean-small grain/red clover (3-yr), and maize-soybean-small grain/alfalfa-alfalfa (4-yr) cropping systems in Boone, IA, averaged over the 2003–2011 study period. Variable means are normalized on a 0 to 1 scale, with 1 representing the cropping system with the largest absolute value for that variable (N = 36 per cropping system). Per-

formance metrics included: maize and soybean yield, rotation-level harvested crop mass, net returns to land and management, manufactured N fertilizer and herbicide application rate, fossil energy use, labor requirements, freshwater toxicity potential and weed seedbank decline (measured as exponential decay constant).

doi:10.1371/journal.pone.0047149.g003

Grain production in the U.S. is dominated by short rotation systems designed to maximize grain yield and profit. These are important goals but represent only a portion of the many ecosystem services that managed lands may provide [18] and that should be considered when evaluating alternative production systems [1], [19]. We believe that these functions are complementary, rather than competing, considerations for agroecosystem design. The results of this study demonstrate that more rotationally diverse cropping systems may be optimized in multiple dimensions, leveraging small agrichemical inputs with biological synergies arising from enhanced diversity of crop species and management tactics.

An example of the synergizing effects of cropping system diversification can be found in weed management in the 3-yr and 4-yr rotations. Weeds were suppressed as effectively in these systems as in the 2-yr rotation, with declining soil seedbanks and negligible weed biomass, yet herbicide inputs in the 3-yr and 4-yr rotation plots were 6 to 10 times lower, and freshwater toxicity 200 times lower, than in the 2-yr rotation. Improved efficiency and environmental sustainability of weed management in the 3-yr and 4-yr rotations resulted from integrating multiple, complementary tactics in an ecological weed management framework [8], [20]. Mounting evidence for unintended effects resulting from heavy reliance on herbicides highlights the need to re-think the role of herbicides in weed management. Non-target impacts of herbicides include reproductive abnormalities and mortality in vertebrates [5], [21]–[23] and potential for diminished non-crop nectar resources for key pollinator species [17], [24], [25]. Herbicide overuse has also resulted in widespread, accelerating evolution of weed genotypes resistant to one or more modes of herbicide action [26], [27]. Our data indicate that, in the context of a cropping system with weed suppressive characteristics, small herbicide inputs may contribute to a diverse suite of tactics that cumulatively provide effective, reliable, and more durable weed management.

The diversity-productivity-stability relationship has long been a key theme in ecology [28], [29]. Recently, it has been applied in the context of bioenergy crop production to describe increases in biomass and ecosystem services, such as C sequestration, associated with increasing species diversity in polycultures of bioenergy feedstock crop species [30]. Our work supports the application of this concept to cropping systems more broadly. Future gains in more diverse systems may depend upon the application of ecological principles surrounding this relationship to cropping system design [31], [32]. Cropping system diversification in this study included both crop species and management practices. In contrast to the 2-yr rotation, with two species, both of the 3-yr and 4-yr rotations included four crop species. In the 4-yr rotation, further temporal diversification was achieved by including a perennial-only crop phase (alfalfa hay) for one quarter of the rotation sequence. Our results showed productivity gains associated with greater diversity in system-level harvested crop mass and maize and soybean seed yields. We also observed increased stability of profit, with similar long-term means, in the 3-yr and 4-yr rotations compared to the 2-yr rotation.

Similar profits were attained through different pathways in the 3-yr and 4-yr rotations and the 2-yr rotation (Fig. 3). Increased labor, information intensive management and ecosystem services arising from increased biological N fixation (via the clover and alfalfa crops) and contrasting crop phenologies and competitive abilities were substituted in 3-yr and 4-yr rotations for the higher inputs of manufactured N, herbicides and energy from fossil fuels driving the 2-yr rotation. Energy use in maize drove differences among the cropping systems, and manufactured N inputs to maize contributed most strongly to energy balances for this crop. The high sensitivity of agricultural energy use to N fertilizer inputs provides a high-priority target for the redesign of cropping systems for increased sustainability.

Reintegration of crop and livestock production, as represented by the forage legumes and manure applications present in the more diverse systems, is not simply another aspect of cropping system diversification. Instead, it embodies an important principle in sustainable agriculture: system boundaries should be drawn to minimize externalities. Animal manure is produced regardless of whether feed grains are shipped to centralized concentrated animal feeding operations, or produced within integrated crop-livestock farming operations. In the former case, the manure may become a waste product and water pollutant if quantities exceed available land area for field application [33], whereas in the latter case, it contributes directly to crop nutrient requirements, improves soil quality, and reduces fossil fuel subsidies associated with grain transport and external N fertilizer inputs [14].

Substantial improvements in the environmental sustainability of agriculture are achievable now, without sacrificing food production or farmer livelihoods. When agrichemical inputs are completely eliminated, yield gaps may exist between conventional and alternative systems [19]. However, such yield gaps may be overcome through the strategic application of very low inputs of agrichemicals in the context of more diverse cropping systems. Although maize is grown less frequently in the 3-yr and 4-yr rotations than in the 2-yr rotation, this will not compromise the ability of such systems to contribute to the global food supply, given the relatively low contribution of maize and soybean production to direct human consumption and the ability of livestock to consume small grains and forages [34]. Through a balanced portfolio approach to agricultural sustainability, cropping system performance can be optimized in multiple dimensions, including food and biomass production, profit, energy use, pest management, and environmental impacts.

MATERIALS AND METHODS

Site Details and Agronomic Management

To investigate the relative performance of conventional and more diverse cropping systems, we conducted a 9-hectare experiment at the Iowa State University Marsden Farm (Figure S1), in Boone County, IA (42°01′ N; 93°47′ W; 333 m above sea level). The experiment site lies within a region of intensive rain-fed maize and soybean production and is surrounded by farms with high levels of productivity. Soils at the site are deep, fertile Mollisols. The experimental cropping system treatments included a conventionally managed 2-yr rotation (maize/soybean) that received agrichemicals at rates comparable to those used on commercial farms in the region, and more diverse cropping systems – a 3-yr rotation (maize/soybean/small grain + red clover green manure) and a 4-yr rotation (maize/soybean/small grain + alfalfa/alfalfa hay) – managed with reduced N fertilizer and herbicide inputs.

The entire site was planted with oat in 2001 and the cropping systems experiment was established in 2002 using a randomized complete block design with each crop phase of each rotation system present every year in four replicate blocks. Plots were 18 m x 85 m and managed with conventional farm machinery. Spring triticale was used as the small grain in 2003–2005, whereas oat was used in 2006–2010. Synthetic fertilizers were applied in the 2-yr rotation at conventional rates based on soil tests. In the 3-yr and 4-yr rotations, composted cattle manure was applied before maize production at a mean dry matter rate of 8.3 Mg ha^{-1} and substantial amounts of N were added through fixation by red clover and alfalfa [35], [36], [37]. Manure and legume N-fixation in the 3-yr and 4-yr rotations were supplemented with synthetic fertilizers based on soil tests, including the late-spring soil nitrate test for maize production [38]. Weed management in the 2-yr rotation was based largely on herbicides applied at conventional rates. In the 3-yr and 4-yr rotations, herbicides were applied in 38-cm-wide bands in maize and soybean and inter-row zones were cultivated; no herbicides were applied in small grain and forage legume crops. Choices of post-emergence herbicides used in each of the systems were made based on the identities, densities, and sizes of weed species observed in the plots. Other details of the farming practices used in the different cropping systems are described in Liebman et al. [39] and in the online SI materials (Text S1). Sampling procedures for determining crop yields, weed biomass and weed seed densities in soil are also described in the online SI materials (Text S1).

Energy and Economic Analyses

Energy inputs were divided into five categories: seed, fertilizer, pesticides, fuel for field operations, and propane and electricity used for drying maize grain after harvest. Data were obtained from logs describing all field operations, material inputs, and crop moisture characteristics for the experimental plots during the study period. Economic analyses measured performance characteristics of whole rotation systems under contrasting management strategies. We evaluated net returns to land and management on a unit land area basis, with land units divided in two equal portions for maize and soybean in the 2-yr rotation; three equal portions for maize, soybean, and small grains with red clover in the 3-yr rotation; and four equal portions for maize, soybean, small grains with alfalfa, and alfalfa in the 4-yr rotation. Net returns to land and management represented returns to a farm operation calculated without accounting for costs of land (e.g., rent or mortgage payments), management time (e.g., marketing), or possible federal subsidies. Data sources and assumptions for the energy and economic analyses are shown in the online SI materials.

Ecotoxicological Calculations

Freshwater ecotoxicity of pesticide use was estimated using the USEtox model [40]–[42]. Characterization factors (CFs) of ecotoxicity potential for active ingredients included transport to freshwater via surface water, soil, and air. CFs were available for eight of ten active ingredients applied in the three rotations. The two active ingredients for which CFs were unavailable are not of particular concern for freshwater ecotoxicity due either to their low toxicity (mesotrione) or low infiltration and persistence in freshwaters (lactofen) [43].

Statistical Analyses

The experiment was arranged in a randomized complete block design, with all entry points of the three crop rotations (i.e. all crops within each of the rotations) represented in four replicate blocks in each year of the study, for a total of 36 plots. Cropping system effects in time series data were analyzed using hierarchical linear mixed effects repeated measures models, modeling temporally correlated errors with an ARMA (auto-regressive moving average) correlation structure in the nlme package of R v.2.14.1 [44], [45]. Fixed effects included cropping system and experimental phase (startup = 2003 to 2005; established = 2006 to 2011), and random effects included replicate block nested within cropping system and year. Partial correlations were estimated using the corpcor package in R v.2.14.1. In contrast to data for quantitative observations (e.g. crop yield or weed biomass) that varied by replicate block and year, data for input variables, such as synthetic fertilizer or herbicides and associated environmental toxicity metrics, did not vary among blocks for a particular rotation entry point in a given year, but did vary among years. Therefore, site-year was treated as the source of experimental replication for these latter variables in our statistical tests for effects of cropping system and experimental phase. This led to contrasting degrees of freedom in reported F-tests for these two data types. Finally, for variables with non-constant variance among cropping systems over time (crop biomass and profit), we used the 'varIdent' variance function within the nlme package to explicitly model differences in variances among cropping systems for these variables within our mixed effects models.

ACKNOWLEDGMENTS

We are grateful to the many students, postdoctoral research associates and technicians who have made the Marsden Farm study possible.

AUTHOR CONTRIBUTIONS

Conceived and designed the experiments: ASD ML. Performed the experiments: CAC AMJ ML. Analyzed the data: ASD JDH CAC AMJ ML. Contributed reagents/materials/analysis tools: ML. Wrote the paper: ASD JDH ML.

REFERENCES

1. 1. Foley JA, Ramankutty N, Brauman KA, Cassidy ES, Gerber JS, et al. (2011) Solutions for a cultivated planet. Nature 478: 337–342. doi: 10.1038/nature10452 View Article PubMed/NCBI Google Scholar.

2. Robertson GP, Swinton SM (2005) Reconciling agricultural productivity and environmental integrity: a grand challenge for agriculture. Front Ecol Environ 3: 38–46. doi: 10.2307/3868443 View Article PubMed/NCBI Google Scholar.

3. Dubrovsky NM, Burow KR, Clark GM, Gronber JM, Hamilton PA, et al.. (2010) The quality of our nation's waters: Nutrients in the nation's streams and groundwater, 1992–2004. Circular 1350. Reston, VA: U.S. Geological Survey. http://pubs.usgs.gov/circ/1350/. Accessed 2012 Jul 10.
4. Gilliom RJ, Barbash JE, Crawford CG, Hamilton PA, Martin JD, et al.. (2006) The quality of our nation's waters: pesticides in the nation's streams and ground water, 1992–2001. Circular 1291. Reston, VA: U.S. Geological Survey. http://pubs.usgs.gov/circ/2005/1291/. Accessed 2012 Jul 10.
5. Relyea RA (2005) The impact of insecticides and herbicides on the biodiversity and productivity of aquatic communities. Ecol Appl 15: 618–627. doi: 10.2307/3868443 View Article PubMed/NCBI Google Scholar.
6. Rohr JR, McCoy KA (2010) A qualitative meta-analysis reveals consistent effects of atrazine on freshwater fish and amphibians. Environ Health Persp 118: 20–32. doi: 10.2307/3868443 View Article PubMed/NCBI Google Scholar.
7. Tilman D, Cassman KG, Matson PA, Naylor R, Polasky S (2002) Agricultural sustainability and intensive production practices. Nature 418: 671–677. doi: 10.1038/nature01014 View Article PubMed/NCBI Google Scholar.
8. Liebman M, Staver CP (2001) Crop diversification for weed management. In: Liebman M, Mohler CL, Staver CP, editors. Ecological management of agricultural weeds. Cambridge: Cambridge University Press. 322–374.
9. Bennett AJ, Bending GD, Chandler D, Hilton S, Mills P (2012) Meeting the demand for crop production: the challenge of yield decline in crops grown in short rotations. Biol Rev 87: 52–71. doi: 10.1111/j.1469-185X.2011.00184.x View Article PubMed/NCBI Google Scholar.
10. Karlen DL, Varvel GE, Bullock DG, Cruse RM (1994) Crop rotations for the 21st century. Adv Agron 53: 1–45. doi: 10.1016/s0065-2113(08)60611-2 View Article PubMed/NCBI Google Scholar.
11. Grigg DB (1974) The agricultural systems of the world: An evolutionary approach. Cambridge: Cambridge University Press. 358 pp.
12. Grigg DB (1989) English agriculture: An historical perspective. Oxford: Basil Blackwell. 320 pp.
13. Magdoff F, Lanyon L, Liebhardt B (1997) Nutrient cycling, transformations, and flows: implications for a more sustainable agriculture. Adv Agron 60: 1–73. doi: 10.1016/s0065-2113(08)60600-8 View Article PubMed/NCBI Google Scholar

14. Naylor R, Steinfeid H, Falcon W, Galloways J, Smil V, et al. (2005) Losing the links between livestock and land. Science 310: 1621–1622. doi: 10.1126/science.1117856 View Article PubMed/NCBI Google Scholar.

15. U.S. Environmental Protection Agency (EPA) (2011) Pesticides industry sales and usage: 2006 and 2007 market estimates. Washington, D.C.: U.S. Environmental Protection Agency. www.epa.gov/opp00001/pestsales/07pestsales/market_estimates2007.pdf. Accessed 2012 Jul 10.

16. Heap I (2012) International survey of herbicide resistant weeds. http://www.weedscience.org. Accessed 2012 Jul 10.

17. Mortensen DA, Egan JF, Maxwell BD, Ryan MR, Smith RG (2012) Navigating a critical juncture for sustainable weed management. BioSci 62: 75–84. doi: 10.1525/bio.2012.62.1.12 View Article PubMed/NCBI Google Scholar

18. Jordan N, Warner KD (2010) Enhancing the multifunctionality of US agriculture. BioSci 60: 60–66. doi: 10.1525/bio.2012.62.1.12 View Article PubMed/NCBI Google Scholar

19. Seufert V, Ramankutty N, Foley JA (2012) Comparing the yields of organic and conventional agriculture. Nature 485: 229–232. doi: 10.1038/nature11069 View Article PubMed/NCBI Google Scholar

20. Liebman M, Davis AS (2009) Managing weeds in organic farming systems: an ecological approach. In: Organic farming: The ecological system. Francis C, editor. Madison: American Society of Agronomy. 173–196.

21. Hayes TB, Collins A, Lee M, Mendoza M, Noriega N, et al. (2002) Hermaphroditic, demasculinized frogs after exposure to the herbicide atrazine at low, ecologically relevant doses. Proc Nat Acad Sci USA 99: 54376–55480. doi: 10.1525/bio.2012.62.1.12 View Article PubMed/NCBI Google Scholar

22. Orton F, Lutz I, Kloas W, Rutledge EJ (2009) Endocrine disrupting effects of herbicides and pentachlorophenol: in vitro and in vivo evidence. Environ Sci Technol 43: 2144–2150. doi: 10.1021/es8028928 View Article PubMed/NCBI Google Scholar

23. Relyea RA (2005) The lethal impact of Roundup on aquatic and terrestrial amphibians. Ecol Appl 15: 1118–1124. doi: 10.1890/04-1291 View Article PubMed/NCBI Google Scholar

24. Egan JF, Mortensen DA (2012) Quantifying vapor drift of dicamba herbicides applied to soybean. Envir Toxicol Chem 31: 1023–1031. doi: 10.1890/04-1291 View Article PubMed/NCBI Google Scholar

25. Freemark K, Boutin C (1995) Impacts of agricultural herbicide use on terrestrial wildlife in temperate landscapes - a review with special reference to North America. Agric Ecosyst Environ 52: 67–91. doi: 10.1890/04-1291 View Article PubMed/NCBI Google Scholar
26. Powles SB, Yu Q (2010) Evolution in action: plants resistant to herbicides. Ann Rev Plant Biol 61: 317–347. doi: 10.1146/annurev-arplant-042809-112119 View Article PubMed/NCBI Google Scholar
27. Tranel PJ, Riggins CW, Bell MS, Hager AG (2011) Herbicide resistances in Amaranthus tuberculatus: A call for new options. J Agric Food Chem 59: 5808–5812. doi: 10.1021/jf103797n View Article PubMed/NCBI Google Scholar
28. Chase JM, Leibold MA (2002) Spatial scale dictates the productivity-biodiversity relationship. Nature 416: 427–430. doi: 10.1038/416427a View Article PubMed/NCBI Google Scholar
29. Tilman D, Reich PB, Knops J, Wedin D, Mielke T, et al. (2001) Diversity and productivity in a long-term grassland experiment. Science 294: 843–845. doi: 10.1126/science.1060391 View Article PubMed/NCBI Google Scholar
30. Tilman D, Hill J, Lehman C (2006) Carbon-negative biofuels from low-input high-diversity grassland biomass. Science 314: 1598–1600. doi: 10.1126/science.1133306 View Article PubMed/NCBI Google Scholar
31. Smith RG, Gross KL, Robertson GP (2008) Effects of crop diversity on agroecosystem function: Crop yield response. Ecosyst 11: 355–366. doi: 10.1007/s10021-008-9124-5 View Article PubMed/NCBI Google Scholar
32. Smith RG, Mortensen DA, Ryan MR (2010) A new hypothesis for the functional role of diversity in mediating resource pools and weed-crop competition in agroecosystems. Weed Res 50: 185–185. doi: 10.1111/j.1365-3180.2009.00745.x View Article PubMed/NCBI Google Scholar
33. Jackson LL, Keeney DR, Gilbert EM (2000) Swine manure management plans in north-central Iowa: Nutrient loading and policy implications. J Soil Water Cons 55: 205–212. doi: 10.1111/j.1365-3180.2009.00745.x View Article PubMed/NCBI Google Scholar
34. Olmstead J (2011) Feeding the world? Twelve years later, U.S. grain exports are up, so too is hunger. Minneapolis, MN: Institute for Agriculture and Trade Policy. http://www.iatp. org/documents/feeding-the-world. Accessed 2012 Jul 10.

35. Heichel GH, Barnes DK, Vance CP, Henjum KI (1984) N2 fixation, and N and dry matter partitioning during a 4-year alfalfa stand. Crop Sci 24: 811–815. doi: 10.1111/j.1365-3180.2009.00745.x View Article PubMed/NCBI Google Scholar
36. Heichel GH, Vance CP, Barnes DK, Henjum KI (1985) Dinitrogen fixation, and N and dry matter distribution during 4-year stands of birdsfoot trefoil and red clover. Crop Sci 25: 101–105. doi: 10.1111/j.1365-3180.2009.00745.x View Article PubMed/NCBI Google Scholar
37. Fox RH, Piekielek WP (1988) Fertilizer N equivalence of alfalfa, birdsfoot trefoil, and red clover for succeeding corn crops. J Prod Agric 1: 313–317. doi: 10.2134/jpa1988.0313 View Article PubMed/NCBI Google Scholar
38. Blackmer AM, Voss RD, Mallarino AP (1997) Nitrogen fertilizer recommendations for corn in Iowa. PM-1714. Ames, IA: Iowa State University Extension and Outreach. http://www.extension.iastate.edu/Publications/PM1714.pdf. Accessed 2012 Jul 10.
39. Liebman M, Gibson LR, Sundberg DN, Heggenstaller AH, Westerman PR, et al. (2008) Agronomic and economic performance characteristics of conventional and low-external-input cropping systems in the central corn belt. Agron. J. 100: 600–610. doi: 10.2134/jpa1988.0313 View Article PubMed/NCBI Google Scholar
40. Berthoud A, Maupu P, Huet C, Poupart A (2011) Assessing freshwater ecotoxicity of agricultural products in life cycle assessment (LCA): a case study of wheat using French agricultural practices databases and USEtox model. Int J Life Cycle Assess 16: 841–847. doi: 10.2134/jpa1988.0313 View Article PubMed/NCBI Google Scholar
41. Hauschild MZ, Huijbregts M, Jolliet O, MacLeod M, Margni M, et al. (2008) Building a model based on scientific consensus for life cycle impact assessment of chemicals: The search for harmony and parsimony. Environ Sci Technol 42: 7032–7037. doi: 10.1021/es703145t View Article PubMed/NCBI Google Scholar
42. Rosenbaum RK, Bachmann TM, Gold LS, Huijbregts MAJ, Jolliet O, et al. (2008) USEtox-the UNEP-SETAC toxicity model: recommended characterisation factors for human toxicity and freshwater ecotoxicity in life cycle impact assessment. Int J Life Cycle Assess 13: 532–546. doi: 10.1007/s11367-008-0038-4 View Article PubMed/NCBI Google Scholar
43. Cornell PMEP (2008) Pesticide management education program. http://pmep.cce. cornell.edu/. Accessed 2012 Jul 10.

44. Crawley MJ (2007) The R Book. West Sussex: John Wiley and Sons. 942 pp.
45. R Development Core Team (2009) R: A language and environment for statistical computing (R Foundation for Statistical Computing, Vienna, Austria).

Chapter 4

YIELD TRENDS ARE INSUFFICIENT TO DOUBLE GLOBAL CROP PRODUCTION BY 2050

Deepak K. Ray, Nathaniel D. Mueller, Paul C. West, Jonathan A. Foley

Institute on the Environment (IonE), University of Minnesota, Saint Paul, Minnesota, United States of America

ABSTRACT

Several studies have shown that global crop production needs to double by 2050 to meet the projected demands from rising population, diet shifts, and increasing biofuels consumption. Boosting crop yields to meet these rising demands, rather than clearing more land for agriculture has been highlighted as a preferred solution to meet this goal. However, we first need to understand how crop yields are changing globally, and whether we are on track to double production by 2050. Using ~2.5 million agricultural statistics, collected for ~13,500 political units across the world, we track four key global crops—maize, rice, wheat, and soybean—that currently produce nearly two-thirds of global agricultural calories. We find that yields in these top four crops are increasing at 1.6%, 1.0%, 0.9%, and 1.3% per year, non-compounding rates, respectively, which is less than the 2.4% per year rate required to double global production by 2050. At these rates global production in these crops would increase by ~67%, ~42%, ~38%, and ~55%, respectively, which is far below what is needed to meet projected demands in 2050. We present detailed maps to identify where rates must be increased to boost crop production and meet rising demands.

INTRODUCTION

The world is experiencing rising demands for crop production, stemming from three key forces: increasing human population, meat and dairy consumption from growing affluence, and biofuel consumption [1]–[5]. By 2050, global agricultural production may need to be increased by 60%–110% to meet these increasing demands [3], [6]–[7] as well as to provide food security to the ~870 million now chronically undernourished [8]. The only peer-reviewed estimate [3] suggests that crop demand may increase by 100%–110% between 2005 and 2050. Numerous authors have suggested that increasing crop yields, rather than clearing more land for food production, is the most sustainable path for food security [2], [4], [9]–[14]. Moreover, crop yield growth has been shown as an effective tool in reducing global poverty and undernourishment, as farmers themselves constitute the vast majority of the poor and the undernourished [15]–[17].

However, several recent studies indicate that yields may no longer be increasing in different regions of the globe [18]–[23]. Yields are no longer improving on 24–39% of our most important cropland areas [23]. Many of these areas are in top crop producing nations, having rising population, increasing affluence, or some combination of these factors [3], [5], [19], [21]–[23]. This may increase difficulty of meeting future crop production goals but key unknowns remain for developing and targeting strategies: how are crop yields changing across the world, where gains in crop yields are able to meet growing demands, and where crop yields are falling behind.

Here we employ ~2.5 million statistics from a newly developed crop yield and area harvested database covering ~13,500 political units globally from 1961 to 2008, focusing on trends in the recent two decades [4], [23]. We determine the rates of yield change in each political unit for the top four global crops: maize, rice, wheat, and soybean. These four crops together produce about two-thirds of current harvested global crop calories [3], [18]. Using these data, we estimate the best-fit linear, non-compounding rates of yield change between 1989 and 2008 for these crops in each of these political units. Yield change is commonly modeled as a linear function of time [24]–[28] and such models have been used to project future crop yields[28]–[30]. We provide local, country, and global-scale rates of recent crop yield changes to determine where the rates of yield increase could double production by 2050, and where they are insufficient. The impact of negative or even slow rates of yield change in these crops could be severe, especially for low-income countries with rapidly rising population. The underlying data, period analyzed, statistical approach, and comparisons of yield projections are described in the Methods section below, with additional details and analysis in Text S1.

Results

The global average rates of yield increase across ~13,500 political units are 1.6%, 1.0%, 0.9%, and 1.3% per year for maize, rice, wheat, and soybean, respectively (Table 1, Figure 1). A ~2.4% per year rate of yield gains (non-compounding) is needed to double crop production by 2050. Current rates are thus not achieving this goal. At current rates only ~67%, ~42%, ~38%, and ~55% increases in maize, rice, wheat and soybean production, respectively, is possible by 2050.

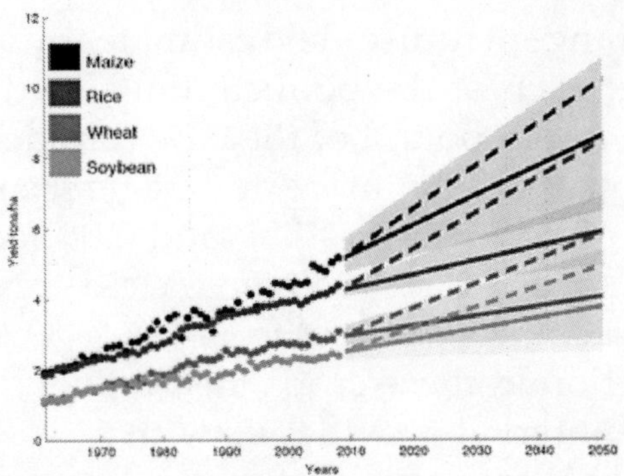

Figure 1. Global projections. Observed area-weighted global yield 1961–2008 shown using closed circles and projections to 2050 using solid lines for maize,

rice, wheat, and soybean. Shading shows the 90% confidence region derived from 99 bootstrapped samples. The dashed line shows the trend of the ~2.4% yield improvement required each year to double production in these crops by 2050 without bringing additional land under cultivation starting in the base year of 2008.

doi:10.1371/journal.pone.0066428.g001

Table 1. Global summary for maize, rice, wheat, and soybean.

	MAIZE	RICE	WHEAT	SOYBEAN
Mean yield change per year (% per year)	1.6	1.0	0.9	1.3
Mean yield change per year (kg/ha/year/year)	84	40	27	31
Projected average yield in 2025 (tons/ha/year)	6.5	4.9	3.6	3.0
Projected average yield in 2050 (tons/ha/year)	8.6	5.9	4.1	3.8
Projected production in 2025 (million tons/year) at fixed crop harvested areas of 2008	1016	760	741	275
Projected production in 2050 (million tons/year) at fixed crop harvested areas of 2008	1343	911	891	347
Projected production shortfall in 2025, as compared to the rate that doubles production by 2050 (million tons/year)	100	160	137	43
Projected production shortfall in 2050, as compared to the rate that doubles production by 2050 (million tons/year)	247	394	388	107
Required extra land (million hectares) to produce the shortfall at 2025 projected yields	15	33	46	14
Required extra land (million hectares) to produce the shortfall at 2050 projected yields	29	67	95	28
Yield in the year 2008 (tons/ha/year)	5.2	4.4	3.1	2.4
90 percent confidence limit in yield change (%/year)	0.8–2.4	0.5–1.4	0.1–1.8	0.3–2.0
90 percent confidence limit in yield change (kg/ha/year/year)	43–124	21–58	4–52	6–58
90 percent confidence limit in production in 2025 (million tons/year) at fixed crop harvested areas of 2008	848–1203	687–846	599–898	214–328
90 percent confidence limit in production in 2050 (million tons/year) at fixed crop harvested areas of 2008	1009–1686	769–1072	418–1182	228–442

As an example consider yields and production in 2025 – the short term – and numbers by 2050 due to current rates of yield change. See Supplementary Data file for yield change rates per country.

doi:10.1371/journal.pone.0066428.t001

doi:10.1371/journal.pone.0066428.t001

We provide a range of future yield estimates by bootstrap sampling crop yield data at each of the political units studied for the period 1989 to 2008. The upper bound of the 90% confidence interval (Table 1, Figure 1) presents a slightly more optimistic scenario, global yields increase at rates of 2.4%, 1.4%, 1.8%, and 2.0% per year for maize, rice, wheat, and soybean, respectively. Yield trends following this upper bound projection could lead to ~101%, ~59%, ~76%, and ~84% increased production in these crops, respectively. The lower bound of our confidence interval provides us with a "worst-case scenario," wherein the global average yield of maize, rice, wheat, and soybean would increase at 0.8%, 0.5%, 0.1%, and 0.3% per year, respectively

(Table 1,Figure 1). At these rates global production could only increase by ~34%, ~21%, ~4%, and ~13% for maize, rice, wheat, and soybean, respectively, by 2050. Further, the yield trajectory diverges, especially for rice and wheat from the 2.4% per year rate (Figure 1). See Figure S1for spatial maps of r^2 at each political unit and statistical diagnostic tests (Text S1 and Figures S2, S3, S4, S5, S6, S7, S8, S9).

In the short term, due to population increases from ~6.7 billion in 2008 to ~8.0 billion in 2025 [5], the 1.6% and 1.3% per year global maize and soybean yield improvements may result in no significant change to the per capita global maize and soybean harvests. However, by 2050 there could be an increase. The much lower rates of rice and wheat yield increases, 1.0% and 0.9% per year, respectively, may result in no change to the per capita rice and wheat harvests to 2050. Thus, if we are to boost the production in these top four global crops that are now responsible for directly providing ~43% of the global dietary energy and ~40% of its daily protein supply [31] from yield increases alone, we have to immediately determine where and exactly by how much yields are changing. To further understand the yield trend patterns, we also track the rates of yield change within ~13,500 political units and report the results at the local and country scales.

Global trends mask the significant variations in the rates of yield change among and within countries (Figure 2). We determine where the within-country yield change rates are ~2.4% per year or above (i.e. doubling rates), where the rates are lower, and where yields are decreasing. We briefly describe these areas, emphasizing areas with doubling and decreasing rates as these areas define the places with the greatest opportunity to meeting growing demand or where to target investments. See Figure S10 for continuous rate (non-categorical) maps in kilograms/hectare/year/year. The influence of observed yields in 2008 on the percent rate of change is described in Figure S11 and related Text S1.

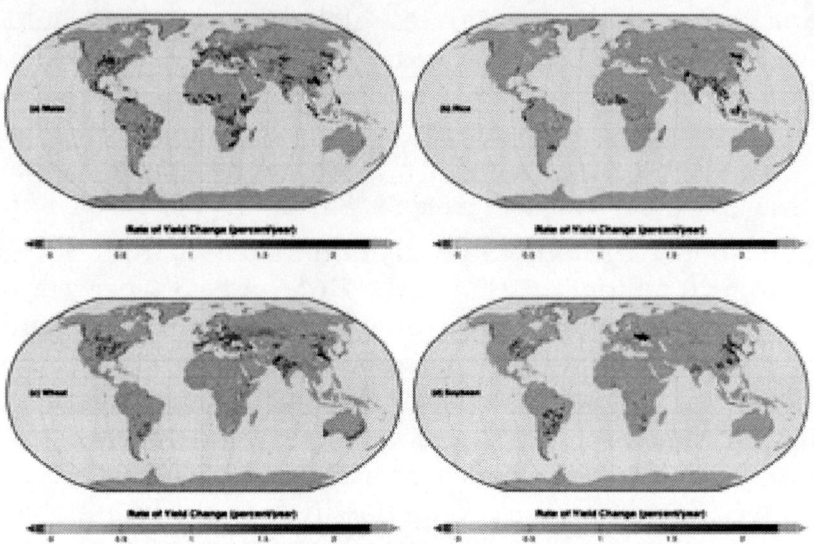

Figure 2. Maps of observed rates of percent yield changes per year. Global map of current percentage rates of changes in (a) maize, (b) rice, (c) wheat, and (d) soybean yields. Red areas show where yields are declining whereas the fluorescent green areas show where rates of yield increase - if sustained - would double production by 2050.

doi:10.1371/journal.pone.0066428.g002

NORTH AND CENTRAL AMERICA

Most of North Dakota and Mississippi, northeastern South Dakota, northwestern Minnesota, and some isolated counties in other United States (U. S.) states are witnessing ~2.4% per year or greater rates of maize yield gains. Similar doubling rates in maize yields are found in the states of Chihuahua, coastal Sinaloa, most of Michoacán and Guanajuato and isolated areas of few other Mexican states as well as El Salvador. Maize yields are decreasing in parts of the U. S. Great Plains states (southern South Dakota, Kansas, Eastern Colorado and parts of southeastern Texas), eastern Mexico (San Luis Potosí, northern Durango, southeastern Coahuila, Nuevo León and Tamaulipas), and in Haiti and Guatemala (Figure 2a). The resultant impact is that the United States has the highest national rates of maize yield improvement in this region of the world (in kg/ha/year^2) followed by Canada, then Cuba, and Mexico (Data S1). In Central American countries such as Honduras and Nicaragua, where maize now provides ~27%, and ~25% of daily dietary energy, respectively [31], and in Panama (~7% of dietary energy), the production gains from their slower 0.5% per year yield improvement rates could be less than those required to keep pace with their population growth (in per capita harvested production terms). In Guatemala, where maize now provides ~36% of dietary energy [31] the yield trends are already negative (−0.7% per year), and as the population is projected to substantially increase [5], a steeper fall in the per capita harvested maize could occur.

Rice yield doubling rates are found only in some isolated areas of North and Central America. The United States has the highest overall rice yield improvement rates (1.2% per year), followed by Mexico (1.1% per year). In Nicaragua and Panama where rice supplies ~16% and ~24% of dietary energy respectively, the per capita rice harvests could fall due to their population growth[5] outpacing their 0.9% and 0.2% per year rice yield improvement rates (Figure 2b). Elsewhere in the Dominican Republic, Costa Rica, and Haiti, where rice provides 16–22% of their daily dietary energy, yields are declining at rates of −0.1% to −0.6% per year. The per capita rice production is likely to increase only in Cuba, where rice yields are increasing 0.9% per year [31]and the population is projected to fall [5].

Wheat yields are increasing at ~2.4% per year or greater only in some counties in the U. S., (mainly in eastern South Dakota, parts of Nebraska, northeastern Kansas, western Mississippi, and Louisiana) (Figure 2c). Wheat yields are decreasing in many parts of the U. S. Great Plains (Montana, western parts of North Dakota, South Dakota, Kansas and Texas and eastern Colorado). In Mexico, areas with doubling rates in wheat yields are observed only in the state of Zacatecas. Nationally, wheat yields in Canada, United States, and Mexico are increasing at 1.3%, 0.8% and 1.1% per year, respectively.

Most areas in the U. S. show increasing soybean yields, with doubling rates in North Dakota, isolated areas of South Dakota, Nebraska, Mississippi, Louisiana, and Georgia (Figure 2d). Soybean yields are decreasing in Kansas, Oklahoma and Texas. Canada and the United States have yields increasing at 0.2% and 1.2% per year, respectively.

SOUTH AMERICA

Most maize areas in South America are achieving doubling rates, with the exception of isolated municípios in Brazil. The overall impact of these varied subnational rates is that maize yields are increasing at 1.7–4% per year in Uruguay, Argentina, Chile, and Brazil and may result in significantly higher per capita maize harvests. Other South American countries such as Venezuela, Peru, Bolivia and Ecuador, where maize provides 2–14% of dietary energy, are achieving yield increases of 1.8–3% per year but due to their population growth [5] may result in no significant changes to their per capita maize harvests at least in the short term.

Rice is grown throughout in South America and yields are improving in most areas (Figure 2b). Rice yield rates are at doubling levels however only in the Cesar and Tolima departments of Colombia, and isolated municípios especially in the states of Pará, Maranhão, and Mato Grosso in Brazil and some areas of Paraguay, Uruguay and Argentina. Decreasing rates of rice yield are found in northeastern Brazil. The overall impact of these subnational rates of rice yield changes is: national rice yields are improving fast in Brazil, Peru, Colombia, Uruguay, and Argentina (1.6–2.7% per year). But due to population growth [5], yield increases alone may be unable to boost the per capita rice harvested in Peru, Argentina and Ecuador in the short term (~2025). Elsewhere in Venezuela and Bolivia (where rice supplies 8–19% of daily dietary energy and rice yields are improving at 1.1–1.4% per year) per capita harvested rice could remain unchanged to 2050. In Suriname, where rice provides ~25% of dietary energy, the very low rates of rice yield improvement, 0.2% per year, may even lead to decreased per capita rice harvests. In Brazil, Uruguay, Guyana, and Paraguay where rice now supplies ~11%, ~7%, ~29%, and ~2% of dietary energy respectively, the per capita rice harvested could increase.

Large extents of doubling wheat yield rates occur only in Argentina and Chile. Wheat yields are decreasing in parts of Peru and in Santiago del Estero in Argentina. The national trends as a consequence: Argentine and Chilean wheat yields are increasing at 1.5% and 1.9% per year respectively and may result in increased per capita wheat harvests (wheat provided ~25% and ~30% of dietary energy, respectively). In Brazil, Colombia, Ecuador, and Uruguay, yields are increasing at 0.1–1.5% per year and may lead to unchanged per capita harvests, (8–31% of dietary energy is now supplied from wheat in these four countries). Per capita wheat harvests could decrease in: Bolivia, Peru, and Paraguay due to lower yield gains of 0.5–1.6% per year.

With the exception of Bolivia and Paraguay, soybean yields are increasing at doubling rates, particularly in many areas of Argentina, Brazil, Venezuela, and Uruguay. Soybean yields are increasing 1.5–2.4% per year in these countries.

EUROPE

Almost everywhere in Europe, except in Moldova, maize yields are increasing but rates of ~2.4% per year are found only in Portugal, the Czech Republic, and Belarus. In Moldova, southern Romania (counties in south and southwest region) and Pomeranian province in Poland maize yields are deceasing and have led national maize yields to change at −4.9% per year in Moldova, 0.7% per year in Romania, and 1.1% per year in Poland. Due to rising maize yields of 0.8–3.0% per year, the per capita harvested maize could increase in many European countries by 2050.

Rice is harvested in only a few European countries and yields are increasing at 0.2–1.5% per year. See Data S1 for the numbers.

Wheat is an important food crop in Europe and harvested in almost all European countries. However, in Eastern Europe (Ukraine, Moldova, southern Romania, Bulgaria, parts of Hungary and Slovak Republic), southern France, and northeastern Spain wheat yields are generally decreasing with the exception of a few regions where yield increases are at doubling rates. Consequently, national wheat yield improvements in European countries are generally <1% per year, with the exception of Estonia (1.5% per year). Even though wheat yield improvements are low, the per capita harvested wheat may increase in some of these European countries because of population declines in Estonia, Germany, Latvia, and Lithuania [5]. Unfortunately, in many other European countries, the low production gains from yield improvement will likely be offset by increasing population, resulting in nearly unchanged per capita wheat harvests. Yields are decreasing in many eastern European countries throughout, where wheat comprises 24–36% of the dietary energy.

Soybean yields are increasing at doubling rates only in small areas in Romania, and in central Italy soybean yields are decreasing. See Data S1 for actual national numbers.

AFRICA

Africa is a continent of contrasts with regards to rates of maize yield change. For example, maize yields are increasing ~2.4% per year in the Nigerian states of Yobe and Adamawa. Similar maize yield improvement rates are found in some other isolated areas of West African nations, Ethiopia, Angola, South Africa, and Madagascar. But maize yields are decreasing in Morocco, Chad, Somalia, Kenya, Zambia, Zimbabwe, Rwanda, Burundi, and Democratic Republic of Congo. Elsewhere, rates of yield improvement are lower than population growth, suggesting that production per capita is likely to decline. These trends are particularly troubling in countries such as Burundi, Chad, Kenya, Morocco, Rwanda, Democratic Republic of Congo, Somalia, Zambia, and Zimbabwe, where yields are decreasing −0.2% to −7.6% per year, population is rising [5], and maize accounts for 5–51% of calorie intake. The only African countries that may witness an increase in per capita maize harvests due to faster maize yield increases are Angola, Ivory Coast, and Mozambique, where yields are increasing at rates of 2.9%, 4.1%, and 3.2% per year, respectively.

In Ivory Coast, Togo and Benin in West Africa, and in Rwanda, rice yield changes are at doubling rates. In contrast, yields are decreasing more than 1% per year in Gambia and 3% per year in Nigeria. Nearly 8% of the dietary energy in Nigeria is supplied from rice [31]. The per capita rice harvests could decrease in almost all the important rice consuming African nations, e.g., Guinea, Madagascar, Mali, Nigeria, and Tanzania, unless yields are boosted further. Only in Ivory Coast there could be an increase on account of its ~2.6% per year yield increases.

Wheat, while grown in only an extremely small area of Africa, though in many countries, is generally increasing yields at high rates. In Angola, Eritrea Malawi, Nigeria, Algeria, Sudan, and South Africa, yields are growing at doubling rates (2.4–3.4% per year).

In Nigeria, and Mpumalanga province of South Africa soybean yields are increasing at doubling rates whereas in Zimbabwe, Democratic Republic of Congo, and Rwanda yields are decreasing.

ASIA AND AUSTRALASIA

Maize yields are generally increasing across Asia. Yield improvement rates are currently on track to double production in some parts of Iran, Pakistan, India, China, Indonesia, Bangladesh, Laos, Cambodia, Vietnam, and Turkey (Figure 2a). In China, Laos, Philippines, Australia, India, Pakistan, and in Turkey, the per capita maize harvested could remain unchanged in the short term but could increase by 2050. In a few countries such as Nepal, Kyrgyzstan, Iraq, Afghanistan, Uzbekistan, and in New Zealand the slower rates of maize yield improvements of 1.3%, 0.3%, 0.9%, 1.1%, 1.6%, and 0.5% per year respectively can result in declines in per capita maize harvests on account of population growth [5]. The per capita maize harvests could increase in Indonesia, Vietnam, Thailand Iran, Myanmar, Bangladesh, Cambodia, Azerbaijan, and in Bhutan [5].

Rice areas with doubling yield rates are found only in some local areas within Afghanistan, India, Bangladesh, Laos, Vietnam and Cambodia. Significant rates of rice yield declines are found in parts of India (especially in parts of Uttar Pradesh, Maharashtra, and Tamil Nadu) and in North Korea. The ~1% per year overall rice yield increase in India could result in no significant change to the overall per capita rice harvested but in China this may remain as only a short term (~2025) problem. Rice provides ~30% and ~27% of the dietary energy in India and China respectively now. On the other hand in the world's third largest rice producer, Indonesia where ~49% of dietary energy is provided by rice, yield improvement rates are much lower at 0.4% per year. The other important Asian rice producers may behave in the following way: no significant change in per capita rice harvests in Pakistan, Nepal, Malaysia, and South Korea; no significant increase in Myanmar, Sri Lanka, Turkey and Bhutan only in the short term, and declines in the Philippines in the long term (~2050). However in Afghanistan, Iraq and Australia due to population growth outpacing production increases from the 2.4%, 0.4%, and 0.3% rates of yield increase per year there could be declines in per capita harvested rice. In North Korea, Kazakhstan, Uzbekistan, and in Turkmenistan yields are declining at −2%, −1.9%, −0.3%, and −1.5% per year respectively. Elsewhere, the per capita rice harvested could increase: Bangladesh, Vietnam, Thailand, Cambodia, Laos, and in Iran due to production increases from high rates of yield improvement (1–2.6% per year) outpacing their population growth [5], and in Japan due to small yield improvement rates (0.5% per year) and small population decreases.

Wheat yields are increasing at doubling rates in parts of Iraq, Iran, and Afghanistan, but only in small parts of the top producing countries of China and India. Wheat yields are decreasing in many areas of India, especially in the states of Madhya Pradesh and Uttaranchal, and in the countries of Kyrgyzstan, and Mongolia, and in the Beijing province of China. Large areas with wheat yield decreases are also found in Queensland, New South Wales, Victoria, and Western Australia, in Australia. The consequence of these wheat yield change rates is diverse. Per capita wheat production could increase in many countries, including China, Iran, and North Korea, because yield increases exceed projected population increases. In contrast, decreases in per capita harvests could occur in: Afghanistan, Georgia, Iraq, Kyrgyzstan, Mongolia, Saudi Arabia, Yemen, and in Australia.

Soybean yields are increasing at ~2.4% throughout China, including the provinces of Jilin, Guangxi, and Guangdong, but are decreasing in Yunnan and Ningxia. Yields are increasing at doubling rates in local areas within Maharashtra in India. In contrast, yields are generally decreasing in the neighboring state of Madhya Pradesh. Doubling rates of soybean yields are also found in Laos and Vietnam, but in North Korea, and Cambodia, yields are decreasing.

DISCUSSION AND CONCLUSIONS

Numerous studies have shown that feeding a more populated and more prosperous world will roughly require a doubling of agricultural production by 2050 [1]–[7], translating to a ~2.4% rate of crop production growth per year. We find that the top four global crops - maize, rice, wheat, and soybean - are currently witnessing average yield improvements only between 0.9 to 1.6 percent per year, far slower than the required rates to double their production by 2050 solely from yield gains. This is because yield improvements are below ~2.4% per year in many areas of our most important agricultural lands. At these rates maize, rice, wheat and soybean production may increase by ~67%, ~42%, ~38%, and ~55% respectively, by 2050 globally. There is a 90% chance that the total global production increase from yields alone would be between 34–101% for maize, 21–59% for rice, 4–76% for wheat, and 13–84% for soybean by ~2050. Thus, if these yield change rates do not increase, land clearing possibly would be needed [3] if global food security is to increase or even maintained (Table 1).

We found that the top three rice and wheat producing nations are witnessing very low yield growth rates. China, India and Indonesia are witnessing rice yield increases of only 0.7%, 1.0%, and 0.4% improvement per year. China, India, and the U. S., the top three wheat producers similarly were witnessing yield increases of only 1.7%, 1.1%, and 0.8% per year, respectively. At these rates we found that yield driven production growth in India and China could result in nearly unchanged per capita rice harvests, but decline steeply in Indonesia.

In many of the smaller crop producing nations, maize, rice, or wheat yield improvement rates are below the 2.4% doubling rate. Unfortunately, a high percentage of total calories consumed in these countries are from these four crops. This is particularly true for maize throughout much of Africa (e.g., Kenya, Zambia, Zimbabwe), Central America (e.g., Guatemala, Nicaragua, Panama), and parts of Asia (e.g., Nepal, Georgia).

Rice provides ~19% of dietary energy globally. Rice provides a higher percentage of total calories consumed in countries such as Dominican Republic, Costa Rica, Haiti, Sierra Leone, Nigeria, and North Korea, yet yields are declining, −0.1% to −3.2% per year. Elsewhere rice yields are increasing too slowly to overcome the impact of their population growth. In some of the world›s top rice producers, e.g. India and China, the per capita production may remain nearly unchanged. In numerous smaller rice producers across the world where rice is an important significant provider of daily dietary energy such as in Peru, Ecuador, Bolivia, Benin, Togo, Myanmar, Philippines, Malaysia, South Korea, Nepal, and in Sri Lanka, the per capita production may also remain unchanged.

Wheat provides ~19% of global dietary energy. Wheat comprises an even larger portion of the diet in some countries where yields are declining, particularly Eastern European countries of Bulgaria, Hungary, Czech Republic, Moldova, Romania, Slovakia, and Ukraine. In many countries, such as Bolivia, Peru, Paraguay, Afghanistan, and Iraq, wheat yield increases are too low to maintain their current per capita harvests.

Our analysis identifies where yield improvements are on track to double production and where investments should be targeted to increase yields. The observed rates of yield change result from several location-specific, socio-economic, and biophysical factors that are described elsewhere [23]. Many studies illustrate that intensification can be unsustainable [32]–[36], but several notable projects in Africa [37] and elsewhere [38] have shown that sustainable intensification is possible and necessary to boost global crop production.

Clearly, the world faces a looming and growing agricultural crisis. Yields are not improving fast enough to keep up with projected demands in 2050. However, opportunities do exist to increase production through more efficient use of current arable lands [4] and increasing yield growth rates by spreading best management practices and closing yield gaps under different management regimes [38]–[42] across the globe. A portion of the production shortfall could also be met by expanding croplands, but at a high environmental cost to biodiversity and carbon emissions [4], [43]–[45]. Alternatively, additional strategies, particularly changing to more plant-based diets and reducing food waste [4], [46]–[48] can reduce the large expected demand growth in food [3], [4].

METHODS

Data

We used annual crop census reports for harvested areas and yield from ~13,500 political units globally covering 20 years from

1989 to 2008 in this analysis though the database itself covers the years 1961 to 2008. The sum total of these census reports for the 20 years was approximately 1.8 million. Data were collected at three political levels/units depending on data availability: country, state/provinces, and county/district/município/department. Data were not available for all political units for each year. Details of the number of years data was available and its source is given in the Table S1. For the political units where data was missing for some years we estimated crop harvested and yield information using the average of the latest five years of reported data and constraining them with the reported numbers from the higher political unit as explained further in Text S1 and previous work [23].

Population data and its projections per country were from the United Nation›s medium variant projections [5]. Crop production was determined using the projected crop yields at current observed rates of yield change and harvested areas fixed at ~2007. Per capita harvested production is the ratio of production to population and a greater than ±10% change from ~2007 is considered as significant either in the short- (2025) or long-term (2050).

Analysis

We linearly regressed 20 years of crop yields at each of the political units to determine the average linear rates of yield improvement over the observed period. Many previous studies have shown that crop yields change linearly and have used linear regression to project future crop yields [24]–[30]. Here we calculate the non-compounding linear percentage rate by solving$a$ in Equation 1; Y is the yield in the year 2008, $2Y$ is the yield in 2050 (after 42 years):$2Y = \frac{a * Y * 42}{100} + Y$ **(1)** This gives a rate of 2.38% per year or approximately 2.4% per year. For reported numbers at the local- to country- to global-scale the linear percentage changes are the observed changes using 2008 yields as the base year. The actual changes are provided in Data S1 for each crop and country.

Details of the method used, sensitivity to the number of years analyzed, as well as alternate regression methods are provided in Figures S6 and S9. The advantage of analyzing at high spatial resolution is that yield rates can be summarized for other unique levels. For example, we summarized the results for the Brazilian Legal Amazon (Figure S12 and Table S2). We compared our global numbers with other reported estimates. These comparisons are provided in Table S3.

Global maps of the coefficient of variation (r^2) for maize, rice, wheat, and soybean when fitted to 20 years of yield information at each political unit analyzed.

ACKNOWLEDGMENTS

We thank George Allez for help in collecting and digitizing a portion of the data, James Gerber for help with Figure 2, and Emily Cassidy for directing us to the Food and Agricultural Organization's Food Balance Sheets. This work also benefitted from the comments of the Foley lab members.

AUTHOR CONTRIBUTIONS

Conceived and designed the experiments: DKR JAF. Performed the experiments: DKR. Analyzed the data: DKR NDM PCW JAF. Contributed reagents/materials/analysis tools: DKR NDM JAF. Wrote the paper: DKR NDM PCW JAF.

REFERENCES

1. Pingali P (2006) Westernization of Asian diets and the transformation of food systems: Implications for research and policy. Food Policy 32: 281–298.
2. Godfray HCJ, Beddington JR, Crute IR, Haddad L, Lawrence D, et al. (2010) Food security: The challenge of feeding 9 billion people. Science 327: 812–818. doi: 10.1126/science.1185383
3. Tilman D, Balzer C, Hill J, Befort BL (2011) Global food demand and the sustainable intensification of agriculture. Proc Natl Acad Sci USA 108: 20260–20264. doi: 10.1073/pnas.1116437108
4. Foley JA, Ramankutty N, Brauman KA, Cassidy ES, Gerber JS, et al. (2011) Solutions for a cultivated planet. Nature 478: 337–342. doi: 10.1038/nature10452

5. United Nations, Department of Economic and Social Affairs. Population Division, Population Estimates and Projections Section Available:http://esa.un.org/unpd/wpp/index.htm.
6. FAO (2009)Global agriculture towards 2050. Rome, FAO.
7. OECD/FAO (2012), OECD-FAO Agricultural Outlook 2012–2021, OECD Publishing and FAO. Available: http://dx.doi.org/10.1787/agr_outlook-2012-en.
8. FAO WFP, IFAD (2012) The State of Food Insecurity in the World 2012. Economic growth is necessary but not sufficient to accelerate reduction of hunger and malnutrition. Rome, FAO.
9. Phalan B, Balmford A, Green RE, Scharlemann JPW (2011) Minimising the harm to biodiversity of producing more food globally. Food Policy 36: S62–S71.
10. Phalan B, Onia M, Balmford A, Green RE (2011) Reconciling food production and biodiversity conservation: Land sharing and land sparing compared. Science 333 (6047) 1289–1291. doi: 10.1126/science.1208742
11. Green RE, Cornell SJ, Scharlemann JPW, Balmford A (2005) Farming and the fate of wild nature. Science 307 (5709) 550–555. doi: 10.1126/science.1106049
12. Matson PA, Vitousek PM (2006) Agricultural intensification: Will land spared from farming be land spared for nature? Conservation Biology 20 (3) 709–710. doi: 10.1111/j.1523-1739.2006.00442.x
13. Tscharntke T, Clough Y, Wanger TC, Jackson L, Motzke I, et al. (2012) Global food security, biodiversity conservation and the future of agricultural intensification. Biological Conservation 151 (1) 53–59. doi: 10.1016/j.biocon.2012.01.068
14. Hulme MF, Vickery JA, Green RE, Phalan B, Chamberlain DE, et al. (2013) Conserving the birds of Uganda's banana-coffee arc: Land sparing and land sharing compared. PLOS ONE 8 (2) doi: 10.1371/journal.pone.0054597.
15. Fan S, Gulati A, Dalafi S (2007) Synthesis: Lesson and Challenges. In The Dragon and the Elephant, Agricultural and rural reforms in China and India, A. Gulati and S. Fan editors, 465–488, The Jon Hopkins University Press.
16. Andrew D, Ephraim C (2011) The Malawi agricultural input subsidy programme: 2005/06 to 2008/09. International Journal of Agricultural Sustainability 9: 1, 232–247. doi: 10.3763/ijas.2010.0567

17. Pingali PL (2012) Green Revolution: Impacts, limits, and the path ahead. Proc Natl Acad Sci USA 109: 12302–12308. doi: 10.1073/pnas.0912953109
18. Cassman KG (1999) Ecological intensification of cereal production systems: Yield potential, soil quality, and precision agriculture. Proc Natl Acad Sci USA 96: 5952–5959. doi: 10.1073/pnas.96.11.5952
19. Finger R (2010) Evidence of slowing yield growth – The example of Swiss cereal yields. Food Policy 35: 175–182. doi: 10.3763/ijas.2010.0567
20. Peltonen-Sainio P, Jauhiainen L, Laurila IP (2009) Cereal yield trends in northern European conditions: Changes in yield potential and its realization. Field Crops Research 110: 85–90. doi: 10.3763/ijas.2010.0567
21. Brisson N, Gate P, Gouache D, Charmet G, Francois-Xavier O, et al. (2010) Why are wheat yields stagnating in Europe? A comprehensive data analysis for France. Field Crops Research 119: 201–212. doi: 10.1016/j.fcr.2010.07.012
22. Lin M, Huybers P (2012) Reckoning wheat yield trends. Environ Res Lett 7 doi:10.1088/1748-9326/7/2/024016.
23. Ray DK, Ramankutty N, Mueller ND, West PC, Foley JA (2012) Recent patterns of crop yield growth and stagnation. Nature Communications 3: 1293 doi:10.1038/ncomms2296.
24. Hafner S (2003) Trends in maize, rice, and wheat yields for 188 nations over the past 40 years: a prevalence of linear growth. Agriculture, Ecosystems and Environment 97: 275–283. doi: 10.1016/j.fcr.2010.07.012
25. Dyson T (1999) World food trends and prospects to 2025. Proc Natl Acad Sci USA 96: 5929–5936. doi: 10.1073/pnas.96.11.5929
26. Dawe D, Dobermann A, Moya P, Abdulrachman S, Singh B, et al. (2000) How widespread are yield declines in long-term rice experiments in Asia? Field Crops Research 66: 175–193. doi: 10.1016/s0378-4290(00)00075-7
27. Ladha JK, Dawe D, Pathak H, Padre AT, Yadav RL, et al. (2003) How extensive are yield declines in long-term rice-wheat experiments in Asia? Field Crops Research 81: 159–180. doi: 10.1016/s0378-4290(00)00075-7
28. Fischer RA, Edmeades GO (2010) Breeding and cereal yield progress. Crop Sci 50: 85–98. doi: 10.1016/s0378-4290(00)00075-7
29. Jaggard KW, Qi A, Ober ES (2010) Possible changes to arable crop yields by 2050. Phil Trans R Soc B 365: 2835–2851. doi: 10.1098/rstb.2010.0153

30. Supit I (1997) Predicting national wheat yields using a crop simulation and trend models. Agricultural and Forest Meteorology 88: 199–214. doi: 10.1016/s0378-4290(00)00075-7

31. FAO (2013) Food balance sheets. Part of FAOSTAT—FAO database for food and agriculture. Rome: Food and agriculture Organization of U. N. accessed February 2013.

32. Foley JA, DeFries R, Asner GP, Barford C, Bonan G, et al. (2005) Global consequences of land use. Science 309 (5734) 570–574. doi: 10.1126/science.1111772

33. Rockström J, Steffen W, Noone K, Persson A, Chapin S, et al. (2009) A safe operating space for humanity. Nature 461 (7263) 472–475. doi: 10.1038/461472a

34. Donald PF, Green RE, Heath MF (2001) Agricultural intensification and the collapse of Europe's farmland bird populations. Proceedings of the Royal Society B: Biological Sciences 268 (1462) 25–29. doi: 10.1098/rspb.2000.1325

35. Heathcote AJ, Filstrup CT, Downing JA (2013) Watershed sediment losses to lakes accelerating despite agricultural soil conservation efforts. PLOS ONE 8 (1) doi: 10.1371/journal.pone.0053554.

36. Eickhout B, Bouwman AF, van Zeijts H (2006) The role of nitrogen in world food production and environmental sustainability. Agriculture, Ecosystems and Environment 116 (1–2) 4–14. doi: 10.1016/j.agee.2006.03.009

37. Pretty J, Toulmin C, Williams S (2011) Sustainable intensification in African agriculture. International Journal of Agricultural Sustainability 9 (1) 5–24. doi: 10.1016/j.agee.2006.03.009

38. Mueller ND, Gerber JS, Johnston M, Ray DK, Ramankutty N, et al. (2012) Closing yield gaps: nutrient and water management to boost crop production. Nature 490: 254–257. doi: 10.1038/nature11420

39. Lobell DB, Cassman KG, Field CB (2009) Crop Yield Gaps: Their importance, magnitudes, and causes. Annu Rev Environ Resour 34: 179–204. doi: 10.1016/j.agee.2006.03.009

40. Dobermann A, Nelson R (2013) Opportunities and solutions for sustainable food production. United Nations sustainable development solutions network - sustainable agriculture and food systems Available: http://unsdsn.org/thematicgroups/tg7/tg7_resources/.

41. Neumann K, Verburg PH, Stehfest E, Müller C (2010) The yield gap of global grain production: A spatial analysis. Agricultural Systems 103: 316–326.

42. Gregory PJ, George TS (2011) Feeding nine billion: the challenge to sustainable crop production. J Exp Bot 62 (15) 5233–5239. doi: 10.1093/jxb/err232

43. West PC, Gibbs HK, Monfreda C, Wagner J, Barford CC, et al. (2010) Trading carbon for food: Global comparison of carbon stocks vs. crop yields on agricultural land. Proc Natl Acad Sci USA 107: 19645–19648. doi: 10.1073/pnas.1011078107

44. Gibbs H (2012) Trading forests for yields in the Peruvian Amazon. Environ Res Lett 7 doi:10.1088/1748-9326/7/1/011007.

45. Hertel TW (2011) The global supply and demand for agricultural land in 2050: A perfect storm in the making? Am J Agr Econ 93 (2) 259–275.

46. Kummu M, de Moel H, Porkka M, Siebert S, Varis O, et al. (2012) Lost food, waste resources: Global food supply chain losses and their impacts on freshwater, cropland, and fertilser use. Science of the Total Environment 438: 477–489. doi: 10.1016/j.scitotenv.2012.08.092

47. Institution of Mechanical Engineers (2013) Global food; waste not, want not. Available:http://www.imeche.org/knowledge/themes/environment/global-food.

48. Gustavsson J, Cederberg C, Sonesson U, van Otterdijk R, Meybeck A (2011) Global food losses and food waste. Food and Agricultural Organization

Chapter 5

INJURY PROFILE SIMULATOR, A QUALITATIVE AGGREGATIVE MODELLING FRAMEWORK TO PREDICT CROP INJURY PROFILE AS A FUNCTION OF CROPPING PRACTICES, AND THE ABIOTIC AND BIOTIC ENVIRONMENT. I. CONCEPTUAL BASES

[1,2]Jean-Noël Aubertot mail, [1,3]Marie-Hélène Robin

[1]Institut National de la Recherche Agronomique, Unité Mixte de Recherche 1248 Agrosystèmes et Agricultures, Gestion des Ressources, Innovations et Ruralités, Castanet-Tolosan, France

[2]Université Toulouse, Institut National Polytechnique de Toulouse, Unité Mixte de Recherche 1248 Agrosystèmes et Agricultures, Gestion des Ressources, Innovations et Ruralités, Castanet-Tolosan, France

[3]Université de Toulouse, Institut National Polytechnique de Toulouse, Ecole d'Ingénieurs de Purpan, Toulouse, France.

ABSTRACT

The limitation of damage caused by pests (plant pathogens, weeds, and animal pests) in any agricultural crop requires integrated management strategies. Although significant efforts have been made to i) develop, and to a lesser extent ii) combine genetic, biological, cultural, physical and chemical control methods in Integrated Pest Management (IPM) strategies (vertical integration), there is a need for tools to help manage Injury Profiles (horizontal integration). Farmers design cropping systems according to their goals, knowledge, cognition and perception of socio-economic and technological drivers as well as their physical, biological, and chemical environment. In return, a given cropping system, in a given production situation will exhibit a unique injury profile, defined as a dynamic vector of the main injuries affecting the crop. This simple description of agroecosystems has been used to develop IPSIM (Injury Profile SIMulator), a modelling framework to predict injury profiles as a function of cropping practices, abiotic and biotic environment. Due to the tremendous complexity of agroecosystems, a simple holistic aggregative approach was chosen instead of attempting to couple detailed models. This paper describes the conceptual bases of IPSIM, an aggregative hierarchical framework and a method to help specify IPSIM for a given crop. A companion paper presents a proof of concept of the proposed approach for a single disease of a major crop (eyespot on wheat). In the future, IPSIM could be used as a tool to help design ex-ante IPM strategies at the field scale if coupled with a damage sub-model, and a multicriteria sub-model that assesses the social, environmental, and economic performances of simulated agroecosystems. In addition, IPSIM could also be used to help make diagnoses on commercial fields. It is important to point out that the presented concepts are not crop- or pest-specific and that IPSIM can be used on any crop.

INTRODUCTION

Third millennium agriculture must reconcile environmental protection and productivity. The world population is projected to reach 8.7–10 billion by 2050 and annual_production will need to increase by 200 million tons by then to meet the projected 470 million ton demand [1]. Several authors attribute the spectacular increase of agricultural production in the second half of the twentieth century to the massive use of products resulting from chemical synthesis [2]; but this intensive production model is nowadays questioned because of public health, agronomic, environmental, and sometimes socio-economic issues. Concepts in crop protection in intensive agricultural production systems changed from destruction of pests (by which we mean, plant pathogens and animal pests in this paper) by the use of pesticides to pest management with techniques based on the improved knowledge of pest dynamics and their natural enemies and the interaction between pests and crops under the influence of Cropping Practices [3]. It is therefore necessary to combine cultural, genetic, biological, physical and chemical control methods to manage pests through Integrated Pest Management (IPM) strategies in order to maintain the pest population levels below those causing economic losses [4].

True IPM is quite different from the practices recommended up to now [5] and is still faced with agronomic and technical difficulties which can curb its development. Its impact on pests is difficult to estimate because of their multiplicity and of their many interactions within agroecosystems. Studies on the effects of alternative control methods mostly concern a major pest (monospecific approach) while farmers have to manage an injury profile in a given field, i.e. a combination of injury levels caused by multiple pests (multi-specific approach) [6]. Similarly, the research has focused on the effect of one (or a few) control method(s), but farmers usually combine several operations (which may have only partial effects) to limit pest development. Each technical operation is likely to modify the sanitary status of a crop [7]. In addition, not only do cultural practices interact with each other, but also, one technique can be detrimental to some pests and favourable to others. Pest populations are characterised by a very high level of diversity and complexity because of multiple interactions within and between populations and with biological, physical, and chemical environments. This complexity is one of the constraints to the implementation of IPM, in addition to others [8]. In order to reduce the reliance of cropping systems on pesticides, it is therefore necessary to develop tools to help the "vertical integration" (combination of several control methods) and the "horizontal integration" (simultaneous management of several pests) of IPM strategies. Dynamics of pest populations can lead to combinations of injuries on a crop which can in turn lead to quantitative or qualitative damage, which usually results in economic losses for farmers and more generally for society as a whole. However, these relationships are not linear and depend on the production situation as shown by several authors [6], [9], [10]. In this paper, we will assume that the production situation is defined by the physical, chemical and biological components, except for the crop, of a given field (or agroecosystem) and its environment, as well as socio-economic drivers that affect farmer's decisions (adapted from [11]). In this definition, the term "environment" refers to the climate and the territory (i.e. landscape and the associated actors) that can directly or indirectly influence the considered field. In a given production situation, a farmer can design several cropping systems according to his goals, his perception of the socio-economic context and his environment, farm organisation, knowledge and his cognition. However, a given cropping system in a given production situation will be assumed to lead to a unique injury profile.

In order to help design cropping systems, modelling is a key tool [12]. However, because of the complexity of agroecosystems, models usually only address a limited part of agroecosystems. Crop models have been developed for decades but do not take into account interactions with pests (e.g. [13], [14]). Epidemiological models sensu lato have been developed to represent pest dynamics, often to help decision making for pesticide treatments. However, these models usually take into account rather poorly the critical effects of cropping practices [15] due to their multiple consequences on the crop-pest-environment dynamics [16]. In addition, the majority of these models address single pests (except for models such as EPIPRE [17], [18]). So far, the only models that consider injury profiles are damage models [19], [20]. However these models do not predict injury profiles but the quantitative damage that they cause.

There is thus a strong need to develop an innovative approach to predict injury profiles as a function of production situations and cropping practices. Because of the complexity of the considered systems [21], and the lack of representation of the effects of cropping practices and their interactions, the linkage of available crop models to epidemiological models seems unlikely to happen when considering multiple pests [3]. Even if a crop model was available, together with epidemiological models for diseases, weeds and animal pests, taking into account the crop status and the effects of cropping practices, attempting to link them would certainly lead to a dead end because of the propagation error phenomenon as well as the large number of parameters and input variables needed. Alternatively, one could consider statistical approaches to cope with the impossibility of addressing these issues when using mechanistic models. However, datasets with observed injury profiles, cropping systems and production situation are scarce and statistical approaches are thus even more unlikely to succeed than mechanistic modelling approaches.

As an alternative, a generic modelling framework, called IPSIM for Injury Profile SIMulator is proposed. It is deliberately simple in the way mechanisms are represented because the system being described, i.e. the agroecosystem, is far too complex for a truly mechanistic representation. It is based on a simple qualitative hierarchical aggregative approach to represent the effects of various factors affecting injury profiles. This paper presents the basic principles of IPSIM, describing its implementation in a software program and providing an example of its specification for a given crop. A companion paper [22] provides a proof of concept of this innovative modelling approach in the field of crop protection for an important disease of wheat.

MATERIALS AND METHODS

Basic Principles of IPSIM

Figure 1 is a schematic representation of an agroecosystem. This figure is the conceptual basis of IPSIM, although its scope is broader than the system directly addressed by IPSIM. According to the farmer's goals, his farm features, his perception of the environment and of the socio-economic context, as well as his knowledge and cognition, he designs cropping systems that will achieve social, economic and environmental performances, as a function of the production situation. These performances will be highly dependent on the injury profile encountered. The term "cropping system" refers here to "a set of management procedures applied to a given, uniformly treated area, which may be a field, part of a field or a group of fields" [23]. This covers many technical operations, for instance, the choice of the crop sequence, cover cropping, cultivar, tillage practices, date and density of sowing, rate of fertilisation and chemical pest control.

The term "system" is used here because these technical choices are inter-dependent [24].

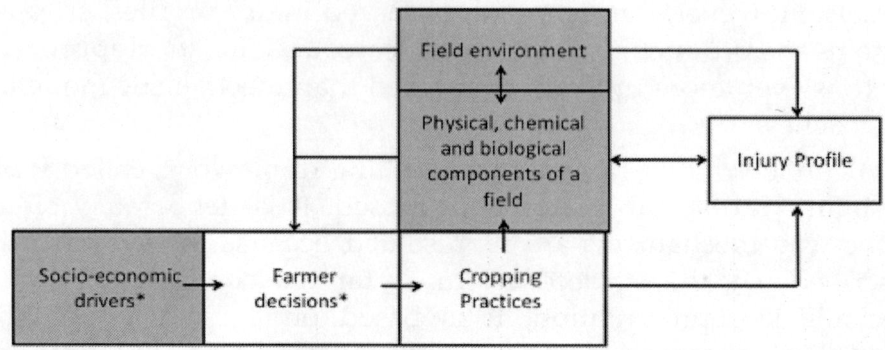

Figure 1 Schematic representation of an agroecosystem and its drivers. In green: components defining the Production Situation (except for the crop). The injury profile is the output variable of IPSIM, whereas its input variables are included within the three following components: cropping practices, field environment, and physical, chemical and biological (crop, pests, beneficials and harmless living organisms) components of the field. *Not taken into account in IPSIM.

doi:10.1371/journal.pone.0073202.g001

IPSIM is embedded in Figure 1, where its output variable is the injury profile. Input variables of IPSIM are embedded within the three following components: cropping practices, field environment, and physical, chemical and biological components of the field (crop, pests, beneficial and harmless living organisms). An injury profile can thus be seen as the result of hierarchical interactions among the cropping practices and the production situation. Qualitative aggregative hierarchical approaches have been used in several fields to help assess the performances of various options when managing a system: industry (e.g. [25], [26]), soil science (e.g. [27]), tourism (e.g. [28], [29]). In the field of agronomy, qualitative aggregative hierarchical models have been used for the assessment of the sustainability of cropping systems ex-ante or ex-post [30]–[32], the assessment of organic systems [33], the management of Genetically Modified crops (e.g. [34]), the assessment of less-favoured areas for agricultural production (e.g. [35]), the evaluation of energy crops for biogas production (e.g. [36]), the assessment of varieties or cultivars (e.g. [37], [38]) and the assessment of the effects of market-gardening cropping systems on soil borne pathogens and animal pests using expert knowledge of advisors [39]. We used this approach to summarise available knowledge in the literature for a given crop and to develop a generic modelling framework for IPM.

IMPLEMENTATION OF IPSIM WITH A SOFTWARE PROGRAM

IPSIM was developed using the DEX method, and is implemented with the DEXi software ([40], http://www-ai.ijs.si/MarkoBohanec/dexi.html). DEX is a method for qualitative hierarchical multi-attribute decision modelling and support based on a breakdown of a complex decision problem into smaller and less complex sub-problems. This tool is generally used to evaluate and analyse decision problems [27], [29], [41].

In this study, it is used for the first time to develop a simulation model that represents the behaviour of an agroecosystem and which quality of prediction can be assessed. The modelling framework has the following features [40]. The sub-problems are hierarchically structured into a tree of attributes that represents the "skeleton" of the model. Terminal nodes of the tree, i.e. leaves or basic attributes, represent input variables of the model (and must be specified by the user). The root node represents the main output: an overall assessment of the evaluated scenarios (an injury profile which is defined by cropping practices and elements of the production situation in this case). The internal nodes of the model are called aggregated attributes. All the attributes in the model are qualitative (ordinal and nominal) rather than quantitative (interval) variables. They take only discrete symbolic values usually represented by words. In the DEX method, the aggregation of values up the tree is defined by "utility functions" based on a set of "if-then" aggregation rules. In our approach, we renamed these functions "aggregating tables" since they are not related to the concept of "utility" in decision theory.

IPSIM Structure

The process of building a DEXi model usually involves the following four steps [40]: (1) identifying the attributes, (2) structuring the attributes, (3) defining attribute scales, and (4) defining the aggregating tables. These steps should be followed for the development of IPSIM using the diagram presented Figure 1. However, only the first three steps can be carried out in a generic way. Only the generic aggregating tables will be described here since most of them are crop-specific.

Structure of the Attributes Used to Predict Injury Profiles

The structure of attributes that predict injury profiles is presented in Figure 2. Each injury can take a limited number of severity levels. For instance, 5 classes (very low, low, medium, high, very high) or 7 classes (nil, very low, low, medium, high, very high and maximum) can be considered in IPSIM. Even if only 10 pests and 5 severity levels are considered for a given crop, a theoretical number of 510 = 9.765625×106 possible injury profiles could thus be simulated with IPSIM. This number is only theoretical since some of these injury profiles are impossible due to interactions among pests. In order to take into account these interactions, IPSIM first calculates the severity for single pests independently, as if one pest only was present (Figure 2). Then, interactions between pests are taken into account according the level of each pest and a simple typology of interaction between two pests: high facilitation, low facilitation, no interaction, low reduction, high reduction (Table 1). Table 1 is used to calculate the overall effect of all other pests on the considered pest. Then, the number of pests with high facilitation, low facilitation, no effect, low reduction, high reduction is calculated (Figure 2) and the overall interactions are calculated according to the aggregating tables presented Table 2. Ultimately, the severity of each pest is calculated using the generic aggregating table presented in Table 3 as a function of the severity that would occur without any other pest, and the overall interactions calculated with the aggregating table presented in Table 2.

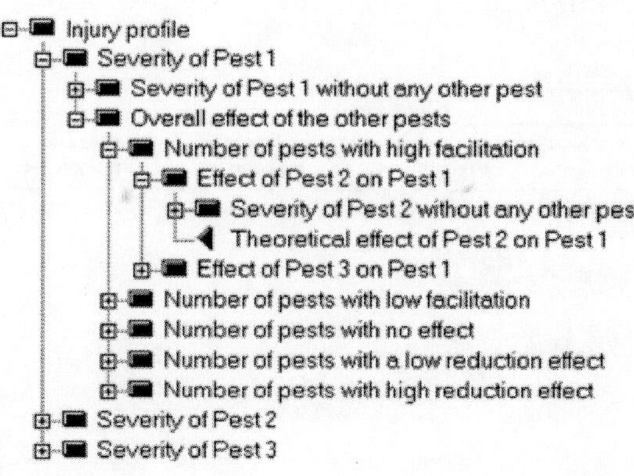

Figure 2 Overall output attributes of IPSIM: description of an injury profile (screenshot of the DEXi software). For the sake of simplicity, only 3 pests are represented in this figure. The severity of a given pest is first calculated independently by IPSIM as if no other pest was present. The aggregated severity of a given pest is then calculated by taking into account the combined effects of all other pests. This is done by considering the theoretical effect of one pest on another according to five levels: high facilitation, low facilitation, no effect, low reduction, high reduction.

doi:10.1371/journal.pone.0073202.g002

Table 1 Generic aggregating table used to represent the effect of one pest on another in IPSIM.

Severity of Pest 2 without any other pest	Theoretical effect of Pest 2 on Pest 1	Actual effect of Pest 2 on Pest 1
Maximum, very high or high	High facilitation	High facilitation
Maximum, very high or high	Low facilitation	Low facilitation
Maximum, very high or high	No effect	No effect
Maximum, very high or high	Low reduction	Low reduction
Maximum, very high or high	High reduction	High reduction
Medium	High and low facilitation	Low facilitation
Medium	No effect	No effect
Medium	High and low reduction	Low reduction
Low or very low	High facilitation	Low facilitation
Low or very low	Low facilitation, no effect, low reduction	No effect
Low or very low	High reduction	Low reduction
Nil	Any	No effect

doi:10.1371/journal.pone.0073202.t001

doi:10.1371/journal.pone.0073202.t

Table 2 Generic aggregating table used to calculate the overall effect on a given pest caused by all the other pests in an injury profile.

Number of pests with high facilitation	Number of pests with low facilitation	Number of pests with no effect	Number of pests with low reduction	Number of pests with high reduction	Overall effects of all other pests
0	0	0	0	0	No effect
0	0	0	0	≥1	High reduction
0	0	0	≥1	0	Low reduction
0	0	0	≥1	≥1	High reduction
0	0	≥1	0	0	No effect
0	0	≥1	0	≥1	High reduction
0	0	≥1	≥1	0	Low reduction
0	0	≥1	≥1	≥1	High reduction
0	≥1	0	0	0	Low facilitation
0	≥1	0	0	≥1	High reduction
0	≥1	0	≥1	0	Low reduction
0	≥1	0	≥1	≥1	Low reduction
0	≥1	≥1	0	0	Low facilitation
0	≥1	≥1	0	≥1	High reduction
0	≥1	≥1	≥1	0	Low reduction
0	≥1	≥1	≥1	≥1	High reduction
≥1	0	0	0	0	High facilitation
≥1	0	0	0	≥1	Low reduction
≥1	0	0	≥1	0	No effect
≥1	0	0	≥1	≥1	Low reduction
≥1	0	≥1	0	0	High facilitation
≥1	0	≥1	0	≥1	Low reduction
≥1	0	≥1	≥1	0	Low reduction
≥1	0	≥1	≥1	≥1	Low reduction
≥1	≥1	0	0	0	High facilitation
≥1	≥1	0	0	≥1	Low reduction
≥1	≥1	0	≥1	0	Low reduction
≥1	≥1	0	≥1	≥1	High reduction
≥1	≥1	≥1	0	0	High facilitation
≥1	≥1	≥1	0	≥1	Low reduction
≥1	≥1	≥1	≥1	0	No effect
≥1	≥1	≥1	≥1	≥1	Low reduction

doi:10.1371/journal.pone.0073202.t002

doi:10.1371/journal.pone.0073202.t002

Table 3 Generic aggregating table used to calculate the severity of one pest in interaction with the other pests of an injury profile.

Severity of the considered pest without any other pests	Overall effect of the other pests	Severity of the considered pest under the influence of other pests
Maximum	High facilitation	Maximum
Maximum	Low facilitation	Maximum
Maximum	No effect	Maximum
Maximum	Low reduction	Very high
Maximum	High reduction	High
Very high	High facilitation	Maximum
Very high	Low facilitation	Maximum
Very high	No effect	Very high
Very high	Low reduction	High
Very high	High reduction	Medium
High	High facilitation	Maximum
High	Low facilitation	Very high
High	No effect	High
High	Low reduction	Medium
High	High reduction	Low
Medium	High facilitation	Very high
Medium	Low facilitation	High
Medium	No effect	Medium
Medium	Low reduction	Low
Medium	High reduction	Very low
Low	High facilitation	High
Low	Low facilitation	Medium
Low	No effect	Low
Low	Low reduction	Very low
Low	High reduction	Very low
Very low	High facilitation	Medium
Very low	Low facilitation	Low
Very low	No effect	Very low
Very low	Low reduction	Very low
Very low	High reduction	Very low
Nil	High facilitation	Nil
Nil	Low facilitation	Nil
Nil	No effect	Nil
Nil	Low reduction	Nil
Nil	High reduction	Nil

doi:10.1371/journal.pone.0073202.t003

doi:10.1371/journal.pone.0073202.t003

Structure of the Attributes Used to Predict the Severity of a Single Pest

The input attributes of IPSIM describe cropping practices, soil and climate (physical and chemical components of the field which partly define the considered production situation), and biological interactions at the territory level (Figure 1). Figure 3 represents the sub-tree used in IPSIM to calculate the severity of a single pest without any interaction with other pests for a given crop. In this sub-tree, cropping practices are composed of cultural, genetic, biological, physical and chemical control actions. Inocula sensu lato are supposed to be non limiting in order to keep basic attributes as simple as possible. The most detailed level is cultural control. It is composed of actions for the management of primary inoculum (through the interaction between crop sequence and tillage for arable crops and prophylactic measures for perennial crops); escape strategies through the choice of the sowing date (some crops are less susceptible to some pests after or before some phenological stages) and mitigation through crop status (as a function of the sowing rate, fertilisation, irrigation, pruning for perennial crops, and application of crop growth regulators). The genetic control represents the level of resistance of the cultivar (or the cultivar mixture) to the considered pest. For some pests, biological control can be applied using living organisms released at the field or greenhouse scale. Physical control consists of using any mechanical, thermal, or electromagnetic actions to limit the pest population. Finally, the attribute "Chemical control" describes the efficacy of pesticide treatments and/or use of non-lethal chemicals such as pheromones or repellents. The effect of soil and climate are described independently and later aggregated in a "Soil and climate" attribute. Finally, the effects of elements (e.g. other fields, hedges, forests) at the territory level are taken into account by describing sources of primary inoculum and beneficials at the territory level, as well as the presence of physical barriers that might limit these interactions between the considered field and its surrounding environment. Harmless living organisms (i.e. neither pests nor beneficials) are not specifically represented in the model.

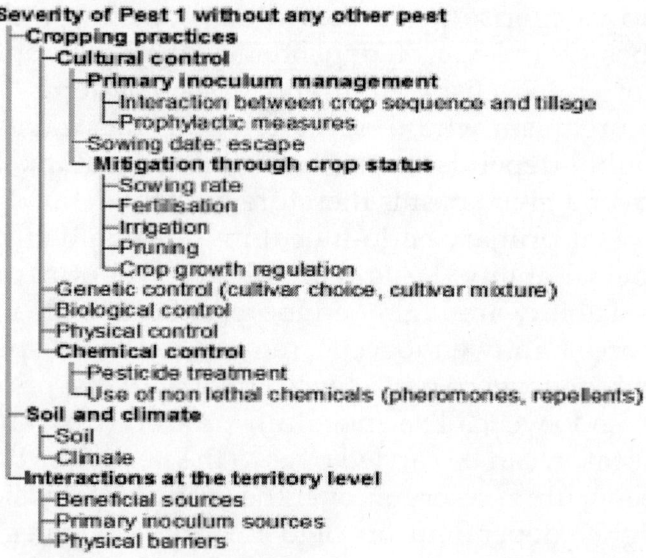

Figure 3 Hierarchical sub-tree to predict the severity of a single pest without any interaction with other pests (screenshot of the DEXi software).

doi:10.1371/journal.pone.0073202.g003

The scales and the aggregating tables used for the attributes presented in Figure 3 cannot be determined in a generic way. They have to be defined according to experimental results, literature, models, or expert knowledge and are specific to the considered crop and pests.

Typology of Simulated Injury Profiles

So far, IPSIM was presented as a simulator of the severity levels for single pests interacting in an injury profile (Figure 2). This detailed information is valuable to researchers, advisers and even farmers to characterise the agronomic performance of cropping practices in a given production situation with regard to potential losses that various pests may cause. However, IPSIM can provide other information, less precise for the injury profile description, but more pertinent for the diagnosis of the overall effects of cropping practices and the biological environment of the considered field on injury dynamics.

We chose to categorise pests according to a simple characteristic that describes their level of dependency to the cropping system: their level of endocyclism (high and low). The term "endocyclic" refers to an organism whose development is mostly restricted to a field and highly depends on the field endo-inoculum. The level of endocyclism of a given pest is therefore directly defined by the level of persistence of primary endo-inoculum sensu lato in a given field and its dispersal ability. Pests with a high level of persistence and low dispersal ability are highly endocyclic. Pests with a low level of persistence are slightly endocyclic, regardless their dispersal ability. Pests with a high level of persistence and a high dispersal ability are moderately endocyclic. The inoculum produced by an endocyclic pest in one season can be carried over to the next, thus building up a cumulative inoculum reservoir over the years. Endocyclic organisms are thus highly dependent on field history. The categorisation of pests into two groups (high/medium and low levels of endocyclism) can help identify the main level to address to control them: the field or territory level.

For example, root-knot nematodes (Meloidogyne spp.) on horticultural crops, wireworms on potato (Agriotes spp.), wheat common bunt (Tilletia spp.), take-all on wheat (Gaeumannomyces graminis var. tritici), dicotyledonous weeds such as Chenopodium album and Fallopia convolvulus are highly endocyclic pests. However, highly endocyclic pests can sometimes be spread to other fields by anthropic activities (e.g. via agricultural machinery, pruning tools, clothes and boots of greenhouse technicians). This dispersal mechanism will not be taken into account in the model. Aphids on several crops (e.g. Brevicoryne brassicae), powdery mildew on grapevine (Erysiphe necator), rusts on cereals (e.g. Puccinia recondita), codling moth on apple tree (Cydia pomonella), and weeds such as some Asteraceae (e.g. Taraxacum dens leonis) or grassy weeds (e.g. Bromus sterilis) are slightly endocyclic pests.

Two aggregating tables were designed to summarise the distribution of final injury levels of single pests using two aggregated variables: the overall final severity of i) highly/moderately and ii) slightly endocyclic pests (Table 4). Considering three levels of final injury (low, medium, high) for each of the two endocyclism groups, a range of nine possible generic injury profiles was proposed (Figure 4) for any agricultural productions worldwide (i.e. major crops; vegetables; vineyard; orchards; horticulture; industrial crops, aromatic and medicinal plants; grassland; in field or in Controlled Environment Agriculture). For production situations where injury profile haves high final injury levels of highly endocyclic pests (IP7, IP8; IP9; Table 5), a better management of primary inoculum production at the field level should be undertaken (e.g. interaction between by crop sequence and tillage; stubble management, volunteer management, stale seedbeds and sanitation measures for perennial crops). For production situations with injury profiles with high levels of slightly endocyclic pests (IP3, IP6; IP9; Table 5), special attention should be paid to i) the management of inoculum production at the territory level (e.g. spatial distribution of cropping systems, management of primary inoculum production in the neighbouring fields or waste piles, management of interstitial spaces to promote beneficials); ii) escape strategies (sowing date adaptation); iii) mitigation through the crop status (e.g. cultivar choice, sowing rate, nitrogen fertilisation, irrigation).

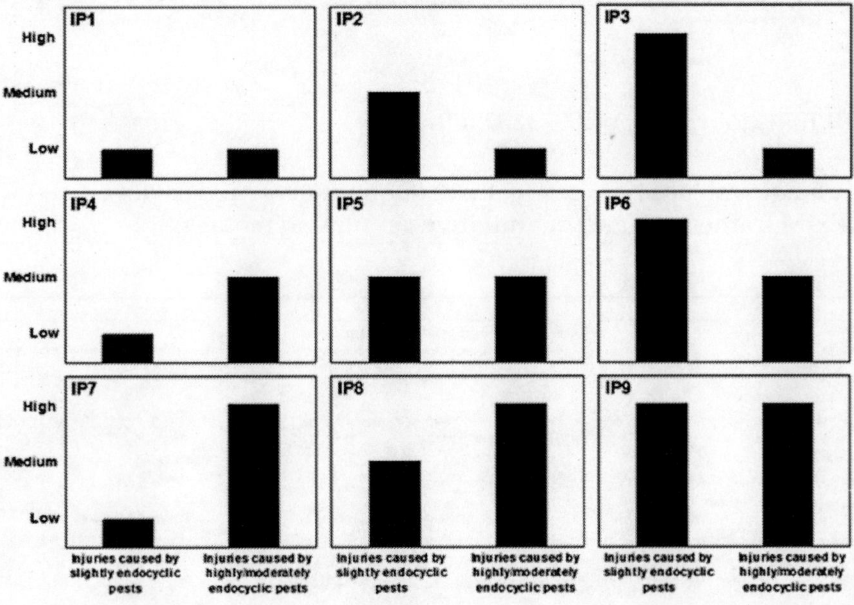

Figure 4 Typology of injuries caused by multiple pests on a crop for given Cropping Practices in a given Production Situation using nine generic Injury Profiles (IP1–IP9).

These Injury Profiles are determined by the final levels of the injuries caused by slightly and highly/moderately endocyclic pests (plant pathogens, weeds and animal pests). They can be used to perform cross-cutting analyses for a wide range of agricultural productions.

doi:10.1371/journal.pone.0073202.g004

Table 4 Generic aggregating table used to define the level of severity of slightly endocyclic pests in an Injury Profile as a function of the final injury level of single pests.

Number of slightly endocyclic pests with a very high or maximum final injury level	Number of slightly endocyclic pests with a low, medium or high final injury level	Number of slightly endocyclic pests with a null or very low final injury level	Overall severity of slightly endocyclic pests in the Injury Profile
>1	>1	>1	High
>1	>1	0	High
>1	0	>1	High
>1	0	0	High
0	>1	>1	Medium
0	>1	0	Medium
0	0	>1	Low
0	0	0	Low

The same aggregating table is used to define the level of severity of highly/moderately endocyclic pests.
doi:10.1371/journal.pone.0073202.t004

doi:10.1371/journal.pone.0073202.t004

Table 5 Equivalence between features of qualitative models developed within the IPSIM framework and quantitative simulation models.

Feature	Qualitative simulation models such as the ones developed with the IPSIM framework	Quantitative simulation models
Type of input variables	Nominal, ordinal, or interval	Interval
Type of state variables	Ordinal	Interval
Type of output variables	Ordinal (can be transformed into static interval or even dynamic interval)	Interval
Model structure	Aggregation tree	Equations
Specification of the model structure	Aggregating tables	Parameters
Analysis of model's behaviour	Table of local and global weights for each input and aggregated attributes	Sensitivity analyses to input variables and parameters
Measures of agreement (non exhaustive)	Proportion of correctly predicted ordinal classes; non parametric Wilcoxon signed rank test to analyse if the distribution of errors is significantly biased or not; matched marginal distribution analysis or joint distribution analysis in a square contingency table	Bias; Mean Absolute Error; Root Mean Squared Error; Efficiency

doi:10.1371/journal.pone.0073202.t005

doi:10.1371/journal.pone.0073202.t005

RESULTS

Implementation of IPSIM Generic Framework into a Simulation Model, an Example

This article aims to present the whole modelling process: i) development of a conceptual framework; ii) implementation of this conceptual scheme into a simulation model for a simple case; iii) simulation to exemplify potential uses of IPSIM models. The specification of IPSIM will be performed for a simple injury profile on wheat: two highly endocyclic diseases (eyespot and sharp eyespot) and a slightly endocyclic disease (brown rust). Eyespot, caused by the necrotrophic and soil-borne fungi Oculimacula yallundae and O. acuformis, anamorph Pseudocercosporella herpotrichoides is considered to be the most important stem-base disease of cereals in temperate countries. In France, sharp eyespot, another soil-borne fungus caused by Rhizoctonia cerealis, is one of the minor diseases of the foot disease complex of winter wheat, but is thought to interact strongly with eyespot. The two pathogens show distinct antagonistic behaviour within the infected stem base, which translates into a negative correlation between sharp eyespot and eyespot incidence [42]–[44]. Finally, brown rust, caused by Puccinia triticina, is the most common rust disease of wheat and is now recognised as an important pathogen in wheat production worldwide, causing significant yield losses over large geographical areas [45]. As opposed to the first two soil-borne diseases which are disseminated over short distances, brown rust is an obligate, airborne disease with conidia which are wind-dispersed over hundreds of kilometres, resulting in rust epidemics on a continental scale [46].

The design of IPSIM-Wheat-Eyespot and the evaluation of its predictive quality is described in a companion paper [22]. For the sake of simplicity and readability, the other two models will not be presented in detail, but their development was similar to the one presented in [22]. The two models for eyespot and sharp eyespot are similar in terms of structure (tree) and aggregating tables because the impact of cropping practices on sharp eyespot is similar to that on eyespot [42]. However, since brown rust is an airborne disease, the effects of primary inoculum management at the field scale are less important than for the two soil-borne diseases. For this airborne disease, the main control methods are: i) mitigation through crop status (using a resistant cultivar for instance) and; ii) the management of primary inoculum sources at the territory level. For this schematic injury profile, we will assume that no direct interactions occur between the two soil-borne, stem-base diseases and this airborne, foliar disease.

Simulation Scenarios

The use of the model presented in the first sub-section of the "results" section is exemplified for three contrasting cropping practices in a given production situation (Figure 5). The three cropping practices considered were: intensive, integrated and organic systems. The intensive system is a wheat monoculture with a high level of inputs and a high-yielding cultivar susceptible to diseases, aiming at a high yield level. The integrated system is characterised by a limited use of inputs, with a lower-yielding cultivar than the former system, but less susceptible to diseases, a short wheat rotation, and a satisfactory yield level. The organic system is characterised by low inputs, with a disease-resistant cultivar with a limited yield, associated with a long wheat rotation and appropriate crop management. The three systems were tested in the same production situation, with a weather scenario favourable to the development of the considered diseases.

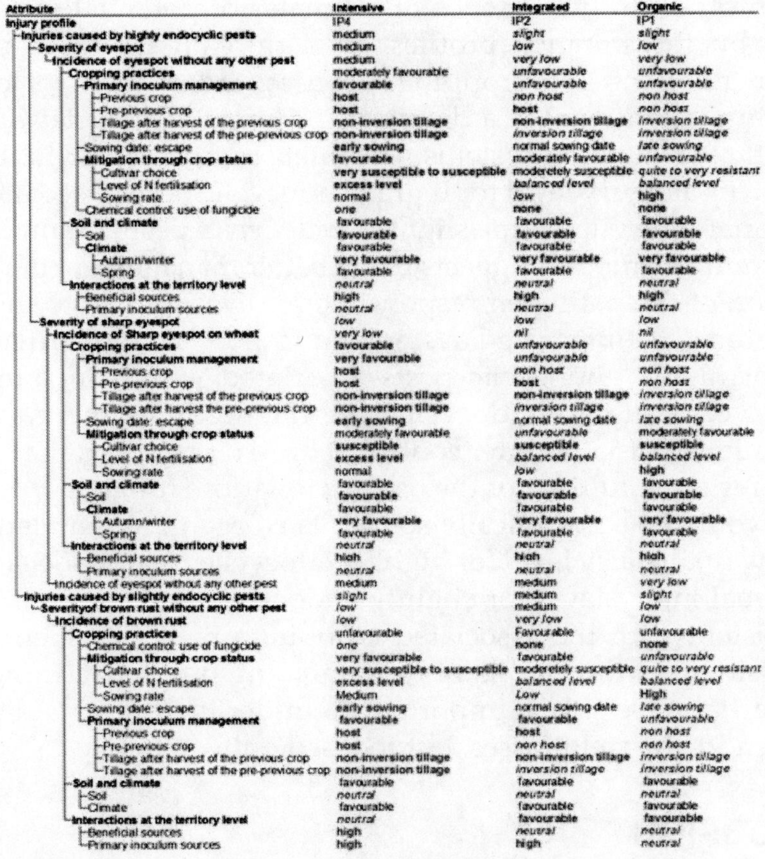

Attribute	Intensive	Integrated	Organic
Injury profile	IP4	IP2	IP1
Injuries caused by highly endocyclic pests	medium	slight	slight
Severity of eyespot	medium	low	low
Incidence of eyespot without any other pest	medium	very low	very low
Cropping practices	moderately favourable	unfavourable	unfavourable
Primary inoculum management	favourable	unfavourable	unfavourable
Previous crop	host	non host	non host
Pre-previous crop	host	host	non host
Tillage after harvest of the previous crop	non-inversion tillage	non-inversion tillage	inversion tillage
Tillage after harvest of the pre-previous crop	non-inversion tillage	inversion tillage	inversion tillage
Sowing date: escape	early sowing	normal sowing date	late sowing
Mitigation through crop status	favourable	moderately favourable	unfavourable
Cultivar choice	very susceptible to susceptible	moderetely susceptible	quite to very resistant
Level of N fertilisation	excess level	balanced level	balanced level
Sowing rate	normal	low	high
Chemical control: use of fungicide	one	none	none
Soil and climate	favourable	favourable	favourable
Soil	favourable	favourable	favourable
Climate	favourable	favourable	favourable
Autumn/winter	very favourable	very favourable	very favourable
Spring	favourable	favourable	favourable
Interactions at the territory level	neutral	neutral	neutral
Beneficial sources	high	high	high
Primary inoculum sources	neutral	neutral	neutral
Severity of sharp eyespot	low	low	low
Incidence of Sharp eyespot on wheat	very low	nil	nil
Cropping practices	favourable	unfavourable	unfavourable
Primary inoculum management	very favourable	unfavourable	unfavourable
Previous crop	host	non host	non host
Pre-previous crop	host	host	non host
Tillage after harvest of the previous crop	non-inversion tillage	non-inversion tillage	inversion tillage
Tillage after harvest the pre-previous crop	non-inversion tillage	inversion tillage	inversion tillage
Sowing date: escape	early sowing	normal sowing date	late sowing
Mitigation through crop status	moderately favourable	unfavourable	moderately favourable
Cultivar choice	susceptible	susceptible	susceptible
Level of N fertilisation	excess level	balanced level	balanced level
Sowing rate	normal	low	high
Soil and climate	favourable	favourable	favourable
Soil	favourable	favourable	favourable
Climate	favourable	favourable	favourable
Autumn/winter	very favourable	very favourable	very favourable
Spring	favourable	favourable	favourable
Interactions at the territory level	neutral	neutral	neutral
Beneficial sources	high	high	high
Primary inoculum sources	neutral	neutral	neutral
Incidence of eyespot without any other pest	medium	medium	very low
Injuries caused by slightly endocyclic pests	slight	medium	slight
Severityof brown rust without any other pest	low	medium	low
Incidence of brown rust	low	very low	low
Cropping practices	unfavourable	Favourable	unfavourable
Chemical control: use of fungicide	one	none	none
Mitigation through crop status	very favourable	favourable	unfavourable
Cultivar choice	very susceptible to susceptible	moderetely susceptible	quite to very resistant
Level of N fertilisation	excess level	balanced level	balanced level
Sowing rate	Medium	Low	High
Sowing date: escape	early sowing	normal sowing date	late sowing
Primary inoculum management	favourable	favourable	unfavourable
Previous crop	host	host	non host
Pre-previous crop	host	non host	non host
Tillage after harvest of the previous crop	non-inversion tillage	non-inversion tillage	inversion tillage
Tillage after harvest of the pre-previous crop	non-inversion tillage	inversion tillage	inversion tillage
Soil and climate	favourable	favourable	favourable
Soil	neutral	neutral	neutral
Climate	favourable	favourable	favourable
Interactions at the territory level	neutral	favourable	favourable
Beneficial sources	high	neutral	neutral
Primary inoculum sources	high	high	neutral

Figure 5 Example of simulation outputs for wheat obtained for three cropping systems (intensive, integrated and organic) in a given production situation (screenshot of the DEXi software).

Three pests in interaction were taken into account in these simulations: eyespot, sharp eyespot and brown rust.

doi:10.1371/journal.pone.0073202.g005

Simulation Results

The DEXi software computed the aggregated attribute values of the model presented in the first sub-section of the "results". In the same production situation, the three cropping practices led to contrasting injury profiles. In the absence of estimates of potential yield losses caused by these injury profiles, it is difficult to provide direct recommendations for cropping practices adaptations. However, these simulations enable a diagnosis in terms of pest development for the three simulated systems. The intensive system led to IP4, i.e. a medium final injury level for highly endocyclic pests associated with a low final injury level for slightly endocyclic pests (Figure 5). For this system, the model suggests that a better management of primary inoculum of the pathogen responsible for eyespot injury should be considered. The integrated system led to IP2, i.e. a low final injury level for highly endocyclic pests associated with a medium final injury level for slightly endocyclic pests (Figure 5). For this system, the model suggests that a better control of brown rust through the use of a more resistant cultivar or the use of a low-dose fungicide, provided that it would be economically sound. The organic system led to IP1, i.e. a low final injury level for highly endocyclic pests associated with a low final injury level for slightly endocyclic pests (Figure 5). This is consistent with the associated cropping practices which aims at minimising pest development by combining prophylactic measures with partial effects. It is important to underline that this diagnosis did not address yield losses, but focused only on injury.

DISCUSSION

Potential Uses of IPSIM Models

These simulations illustrate how IPSIM can be used to assess ex-ante the performance of various cropping systems with regard to the control of pest injury on a given crop. This information is useful when designing innovative cropping systems, either by prototyping, e.g. [47], simulation, e.g. [48], or expert knowledge, e.g. [12]. Since climate significantly affects injury profiles, weather frequency analyses are needed, using a set of input variables describing a wide range of climatic scenarios so that the information provided by IPSIM is robust in the face of weather variability. However, IPSIM cannot be seen as a model to design innovative cropping systems in silico for two major reasons. First, crop damage is not simulated by IPSIM, which makes it difficult to rank pests with respect to the crop losses they cause. Second, the social, economic and environmental performance of the simulated cropping systems are not calculated. To tackle this problem, IPSIM could be coupled to a damage model (such as RICEPEST [6], [19], [20] or WHEATPEST [20]) that would predict yield losses as a function of the injury profiles encountered and other relevant variables. Alternatively, a crop model (e.g. STICS [14]) could be used, with a set of single damage functions (such as the ones used in WHEATPEST [20]), and coupled with IPSIM. Then, once the damage caused by a given injury profile in a given production situation has been predicted, a more general framework, such as MASC, [30] or DEXiPM, [49], could be used to predict the social, economic and environmental performance of the tested systems in a given production situation. This approach will help design innovative cropping systems less vulnerable to pests. Using that modelling framework, IPSIM would be the missing link to fill the gap between crop models that can help predict performance of pest-free cropping systems and epidemiological models that generally do not represent the effects of crop status under the influence of cropping practices. In addition, models developed with IPSIM could be used to create typologies of injury profiles at a regional, national, continental or even worldwide scale, using a schematic description of soil and climate, together with a description of the diversity of cropping practices. This should reveal the main injury profiles encountered and help design strategies to control them with better vertical and horizontal integration of IPM. If the corresponding damage models were available, the typology produced could help prioritise objectively research efforts on the main harmful pests.

IPSIM could also be used in an ex-post analysis to understand the behaviour of commercial field or experimental plots. Finally, it can be viewed as a communication tool for groups, as well as to teach practitioners and students. Knowledge of several scientific fields involved in crop protection, as well as several types of expertise (of scientists, extension engineers, or farmers) can be built into IPSIM, offering a framework for these various communities to interact and combine their knowledge.

LIMITATIONS OF THE APPROACH

Like any other model, the predictive quality of IPSIM should be assessed prior to its use. This highlights the urgent need to collect data in commercial fields describing the input and output variables of IPSIM (i.e. cropping practices, soil and climate, field environment, injury profiles), along with the social, economic and environmental performances of the monitored agroecosystems. It is important to also add measurements of state variables characterising the crop status (e.g. biomass per area unit, Leaf Area Index) in order to better describe important state variables of the agroecosystem for other possible future analyses of the created datasets. However, due to the lack of datasets containing a description of injury profiles, the confidence that users may have in IPSIM models could also be enhanced by comparing simulation outputs with their own expertise to identify any mismatches. All the information contained within IPSIM models is held in the hierarchical trees and the associated aggregating tables. One of the consequences of this specificity of models developed with the IPSIM framework is that, once developed, the predictive quality of the models can be enhanced easily using experimental datasets by modifying aggregating tables, and, if need be, the structure of the model.

The possible injury profiles that IPSIM models can simulate are numerous. However, observations tend to show that the diversity of injury profiles encountered in commercial fields is much less than the structure of IPSIM models can generate. This results from two mechanisms. First, pests can interact directly (through facilitation, predation, competition for the same ecological niche) or indirectly (though modification of the biotope). This implies that not all potential theoretical injury levels could occur simultaneously. This is a limitation of IPSIM which does not account for the impact of injuries on crop growth. Secondly, the soil, climate, cropping practices and landscape occurring in a given territory might not be diverse enough to lead to all theoretical injury levels (for instance, the theoretical injury profile with all the forms of injury at their maximum level does not exist in reality). Another limitation of IPSIM is the way that interactions between pests are represented. If n pests are considered, n (n-1) interactions are to be described. This is similar to the three-body (or n-body) problem in physics, which has a global analytical solution in the form of convergent power series [50], but that has to be approximated in practice because they converge too slowly. IPSIM models approximate interactions among injuries by arbitrarily calculating the global interaction that would occur between a given injury and the rest of the injury profile defined as the sum of single injuries simulated without taking into account interactions among pests. However, this approximation certainly appears negligible as compared to other necessary simplification hypotheses.

From the conceptual viewpoint, it could be asked why the crop which is entered in the field biological component (Figure 1) does not appear at the first level of the IPSIM tree. After all, pests only experience physical, chemical and biological interactions within agroecosystems and a description of i) the crop status, ii) soil and climate, and iii) the neighbouring environment of the field are indeed the true drivers of pest dynamics. This option was tried when developing IPSIM structure, but led to too complicated a structure, the effect of single cultural operations being overlooked among the numerous levels of the tree. In addition, datasets with a description of cropping practices and injury profiles are extremely scarce. The requirement of additional variables describing the crop status (e.g. in terms of phenology, architecture, biomass, Leaf Area Index) would also lead to greater difficulties in developing IPSIM models and in evaluating its predictive quality.

We recommend to develop models with no more than 7 final injury levels for a single pest. The lack of precision of IPSIM models could be seen as a drawback as compared to quantitative epidemiological models. Firstly this is because these latter models address a much simpler system: a single pest, rather than an injury profile. Secondly, when developing models of complex systems, accuracy should be sought rather than precision. Searching for better precision would certainly lead to an increase in the model's complexity and possibly to a dead end. We believe that the proposed precision of the models that will be developed with the IPSIM framework is more than enough for the main ultimate purpose of the model: helping the design of innovative cropping systems less vulnerable to pests.

POINTS FOR REFLECTION

The presented structure of IPSIM is not exhaustive in terms of control methods that can be undertaken. However, developers of models within the IPSIM framework can always easily modify its structure in order to take into account the effects of control measures not present in Figure 3. For instance, the effect of cultivar mixtures or intercrops could be implemented, provided that the required knowledge is available.

The main breakthrough of IPSIM is to be able to handle complexity in a simple way. Input variables of the IPSIM models should be simple to provide. Most of these input variables will be static variables, except for weather variables that will be dynamic. The price to pay to handle the level of ecological complexity (such as defined by Li [51]) addressed by IPSIM is that IPSIM models are static. This is certainly not a problem to predict the consequences of technical options in a given production situation, but could hamper the linkage with dynamic models as suggested earlier. This limitation could easily be overcome by associating the level of final injury predicted by IPSIM models with generic dynamics. In order to do so, exponential, monomolecular, logistic, Gompertz, or Richards models [52] could be used with generic parameters chosen to represent the qualitative ordinal different injury levels predicted by IPSIM models.

The choice of qualitative variables to describe agroecosystems is relevant for several reasons. Firstly, farmers generally rely on a qualitative perception of their environment to make decisions. This suits the formalism of IPSIM. Secondly, because of the complexity of the system, few datasets are available to describe its components, i.e. the production situation, cropping practices and the injury profile. Using qualitative variables enables one to gather and use various existing datasets that were not acquired for the development of IPSIM models. For instance, datasets from diagnoses of commercial fields or even from experiments may not have used the same severity scale for a given disease. The use of qualitative classes allows data from different origins and/or with different precision to be combined. It is possible to associate interval classes with qualitative attributes. For instance, the 7 levels "nil"; "very low"; "low"; "medium"; "high"; "very high"; and "maximum" can be transformed into [0]; [0–20];

[20]–[40]; [40–60]; [60–80]; [80–100]; [100] intervals of percentage of diseased foliage respectively, if one wants to compare these outputs with observed severities of a disease for instance. Thus, data acquired on various scales can still be combined to strengthen the dataset used to estimate the predictive quality of the model or to improve the aggregating tables.

Table 5 presents the equivalence between features of models developed within the IPSIM framework and more common quantitative simulation models. Input attributes of IPSIM models can be nominal, ordinal or interval variables, unlike quantitative simulation models, which require only interval input variables. The state variables (aggregated attributes) of IPSIM models, including output variables, are ordinal. However, if need be, output variables of IPSIM models can be transformed into interval variables. This transformation can be performed by associating each possible ordinal value with a quantitative value (e.g. static final value of an injury level, or quantitative intervals) or with an injury dynamic. The relationship between variables is described by a tree in aggregative qualitative models, whereas quantitative models use equations. The DEXi software [40] provides a table with the respective weights of input and aggregated attributes on the value of the root node (main output). This table can be seen as an equivalent to a simple sensitivity analysis to input variables for quantitative models, prior to more detailed ones [53]. It is notable that IPSIM models have no parameters. The equivalents of parameters that specify relationships among variables in quantitative models are the aggregating tables. The proportion of situations correctly simulated is a criterion that can be used to characterise the agreement between values simulated with an IPSIM model and observations. In addition, a non-parametric Wilcoxon signed rank test can be used to analyse whether the distribution of errors is significantly biased or not. These criteria can be seen as equivalent to common statistical criteria for quantitative models (Bias, Mean Absolute Error; Root Mean Square Error; Efficiency [54]). At last, methods specific to matched-pairs data with ordered categories can be used. In order to do so, various models comparing matched marginal distributions or analysing the joint distribution in a square contingency table can be applied [55].

The qualitative attributes of IPSIM models can lead to threshold effects. In order to cope with this limitation, a tool, named proDEX was developed to model uncertain expert knowledge [56]. This software offers the definition of probabilistic aggregating tables, where each combination of descendants' values maps to a probability distribution of the aggregated attributes, rather than a single value. In this approach, input values must be categorised prior to their use in the model. Since this process is time-consuming, proDEX allows categorisations to be part of the model definition and the inputs to be entered as interval variables. In combination with probabilistic aggregating tables, categorisations can be made to transform numerical values into probabilistic distributions, eliminating the problem of crisp interval boundaries. Eventually, the proDEX method could permit a useful extension of the modelling approach presented in this paper.

Finally, a website giving online access to all the functionalities of IPSIM is planned. This website will enable researchers, advisors, farmers and students to develop their own models for a wide range of crops.

CONCLUSION

We believe that IPSIM is a useful innovative modelling framework to help vertical and horizontal integrations for IPM. Its output attributes include nine generic injury profiles that are based on a two-level categorisation of the degree of endocyclism of harmful organisms. These nine injury profiles can be seen as a tool to perform cross-cutting typologies of agroecosytems for various types of crop (arable crops, vegetables, orchards, vineyards, Controlled Environment Agriculture), with regard to the main pests that have to be managed. IPSIM will generate new knowledge by combining various sources of information from experiments, diagnoses of commercial field, models, and expert panels in a simple way, despite the high ecological complexity of the system addressed. The associated companion paper provides a proof of concept of the proposed method for a single pest.

AUTHOR CONTRIBUTIONS

Analyzed the data: JNA MHR. Contributed reagents/materials/analysis tools: JNA MHR. Wrote the paper: JNA MHR.

REFERENCES

1. Fess TL, Kotcon JB, Benedito VA (2011) Crop breeding for low input agriculture: a sustainable response to feed a growing world population. Sustainability 3: 1742–1772. doi: 10.3390/su3101742 View Article PubMed/NCBI Google Scholar
2. Kropff MJ, Bouma J, Jones JW (2001) Systems approaches for the design of sustainable agro-ecosystems. Agricultural Systems 70: 369–393. doi: 10.3390/su3101742 View Article PubMed/NCBI Google Scholar
3. Kropff MJ, Teng PS, Rabbinge R (1995) The challenge of linking pest and crop models. Agricultural Systems 49: 413–434. doi: 10.3390/su3101742 View Article PubMed/NCBI Google Scholar
4. Birch ANE, Begg GS, Squire GR (2011) How agro-ecological research helps to address food security issues under new IPM and pesticide reduction policies for global crop production systems. Journal of Experimental Botany 62: 3251–3261. doi: 10.1093/jxb/err064 View Article PubMed/NCBI Google Scholar
5. Ferron P, Deguine JP (2005) Crop protection, biological control, habitat management and integrated farming. A review. Agronomy for Sustainable Development 25: 17–24. doi: 10.1007/978-90-481-2666-8_29 View Article PubMed/NCBI Google Scholar
6. Savary S, Willocquet L, Elazegui FA, Teng PS, Pham Van D, et al. (2000) Rice pest constraints in tropical Asia: characterization of injury profiles in relation to production situations. Plant Disease 84: 341–337. doi: 10.1007/978-90-481-2666-8_29 View Article PubMed/NCBI Google Scholar
7. Zadoks JC (1993) Modern crop protection: developments and perspectives; Zadoks JC, editor. Wageningen Netherlands: Wageningen Pers. ix +309 p.
8. Jeger MJ (2000) Bottlenecks in IPM. Crop Protection 19: 787–792. doi: 10.1007/978-90-481-2666-8_29 View Article PubMed/NCBI Google Scholar

9. Zadoks JC (1984) A quarter century of disease warning, 1958–1983. Plant Disease 68: 352–355. doi: 10.1094/pd-69-352 View Article PubMed/NCBI Google Scholar
10. Daamen RA, Wijnands FG, Vliet Gvd (1989) Epidemics of diseases and pests of winter wheat at different levels of agrochemical input. A study on the possibilities for designing an integrated cropping system. Journal of Phytopathology 125: 305–319. doi: 10.1094/pd-69-352 View Article PubMed/NCBI Google Scholar
11. Breman H, de Wit CT (1983) Rangeland Productivity and Exploitation in the Sahel. Science 221: 1341–1347. doi: 10.1126/science.221.4618.1341 View Article PubMed/NCBI Google Scholar
12. Debaeke P, Munier-Jolain N, Bertrand M, Guichard L, Nolot JM, et al. (2009) Iterative design and evaluation of rule-based cropping systems: methodology and case studies. A review. Agronomy for Sustainable Development 29: 73–86. doi: 10.1094/pd-69-352 View Article PubMed/NCBI Google Scholar
13. Stockle CO, Martin SA, Campbell GS (1994) CropSyst, a cropping systems simulation model: water/nitrogen budgets and crop yield. Agricultural Systems 46: 335–359. doi: 10.1094/pd-69-352 View Article PubMed/NCBI Google Scholar
14. Brisson N, Mary B, Ripoche D, Jeuffroy MH, Ruget F, et al. (1998) STICS: a generic model for the simulation of crops and their water and nitrogen balances. I. Theory and parameterization applied to wheat and corn. Agronomie 18: 311–346. doi: 10.1051/agro:19980501 View Article PubMed/NCBI Google Scholar
15. Aubertot JN, Salam MU, Diggle AJ, Dakowska S, Jedryczka M (2006) SimMat, a new dynamic module of Blackleg Sporacle for the prediction of pseudothecial maturation of L. maculans/L. biglobosa species complex. Parameterisation and evaluation under Polish conditions. In: Koopmann B, Cook S, Evans N, Ulber B, editors. Bulletin OILB/SROP. 277–285.
16. Bergez JE, Colbach N, Crespo O, Garcia F, Jeuffroy MH, et al. (2010) Designing crop management systems by simulation. European Journal of Agronomy 32: 3–9. doi: 10.1051/agro:19980501 View Article PubMed/NCBI Google Scholar
17. Rabbinge R, Rijsdijk FH (1983) EPIPRE: a disease and pest management system for winter wheat, taking account of micrometeorological factors. Bulletin, OEPP 13: 297–305. doi: 10.1051/agro:19980501 View Article PubMed/NCBI Google Scholar

18. Zadoks JC (1988) EPIPRE: research, development and application of an integrated pest and disease management system for wheat. Bulletin SROP. 82–90.

19. Willocquet L, Savary S, Fernandez L, Elazegui FA, Castilla N, et al. (2002) Structure and validation of RICEPEST, a production situation-driven, crop growth model simulating rice yield response to multiple pest injuries for tropical Asia. Ecological Modelling 153: 247–268. doi: 10.1016/s0304-3800(02)00014-5 View Article PubMed/NCBI Google Scholar

20. Willocquet L, Aubertot JN, Lebard S, Robert C, Lannou C, et al. (2008) Simulating multiple pest damage in varying winter wheat production situations. Field Crops Research 107: 12–28. doi: 10.1016/s0304-3800(02)00014-5 View Article PubMed/NCBI Google Scholar

21. Savary S, Mille B, Rolland B, Lucas P (2006) Patterns and management of crop multiple pathosystems. European Journal of Plant Pathology 115: 123–138. doi: 10.1016/s0304-3800(02)00014-5 View Article PubMed/NCBI Google Scholar

22. Robin MH, Colbach N, Lucas P, Monfort F, Cholez C, et al. (2013) Injury Profile SIMulator, a hierarchical aggregative modelling framework to predict an injury profile as a function of cropping practices, and the abiotic and biotic environment. II. Proof of concept: design of IPSIM-Wheat-Eyespot. PLoS ONE. In press.

23. Sebillotte M (1990) Systèmes de culture, un concept opératoire pour les agronomes, L. Combe, D. Picard, Editors, Les Systèmes de Culture, INRA, Paris (1990), 165–196.

24. Meynard JM, Dore T, Lucas P (2003) Agronomic approach: cropping systems and plant diseases. Comptes Rendus Biologies 326: 37–46. doi: 10.1016/S1631-0691(03)00006-4 View Article PubMed/NCBI Google Scholar

25. Oblak L, Novak B, Lipuscek I, Kropivsek J (2007) Launching work orders into production of wood enterprise with the multi-criteria decision-making method. Zbornik Gozdarstva in Lesarstva: 33–39.

26. Rozman C, Pazek K, Bavec F, Bavec M, Turk J, et al. (2006) A multi-criteria analysis of spelt food processing alternatives on small organic farms. Journal of Sustainable Agriculture 28: 159–179. doi: 10.1300/j064v28n02_12 View Article PubMed/NCBI Google Scholar

27. Griffiths BS, Ball BC, Daniell TJ, Hallett PD, Neilson R, et al. (2010) Integrating soil quality changes to arable agricultural systems following organic matter addition, or adoption of a ley-arable rotation. Applied Soil Ecology 46: 43–53. doi: 10.1300/j064v28n02_12 View Article PubMed/NCBI Google Scholar

28. Rozman C, Potocnik M, Pazek K, Borec A, Majkovic D, et al. (2009) A multi-criteria assessment of tourist farm service quality. Tourism Management 30: 629–637. doi: 10.1300/j064v28n02_12 View Article PubMed/NCBI Google Scholar
29. Ars MS, Bohanec M (2010) Towards the ecotourism: a decision support model for the assessment of sustainability of mountain huts in the Alps. Journal of Environmental Management 91: 2554–2564. doi: 10.1016/j.jenvman.2010.07.006 View Article PubMed/NCBI Google Scholar
30. Sadok W, Angevin F, Bergez JE, Bockstaller C, Colomb B, et al. (2009) MASC, a qualitative multi-attribute decision model for ex ante assessment of the sustainability of cropping systems. Agronomy for Sustainable Development 29: 447–461. doi: 10.1007/978-90-481-2666-8_46 View Article PubMed/NCBI Google Scholar
31. Pelzer E, Fortino G, Bockstaller C, Angevin F, Lamine C, et al. (2012) Assessing innovative cropping systems with DEXiPM, a qualitative multi-criteria assessment tool derived from DEXi. Ecological Indicators 18: 171–182. doi: 10.1007/978-90-481-2666-8_46 View Article PubMed/NCBI Google Scholar
32. Pazek K, Rozman C, Bavec F, Borec A, Bavec M (2010) A multi-criteria decision analysis framework tool for the selection of farm business models on organic mountain farms. Journal of Sustainable Agriculture 34: 778–799. doi: 10.1007/978-90-481-2666-8_46 View Article PubMed/NCBI Google Scholar
33. Pazek K, Rozman C (2007) The decision support system for supplementary activities on organic farms. Agricultura (Slovenia) 5: 15–20. doi: 10.1007/978-90-481-2666-8_46 View Article PubMed/NCBI Google Scholar
34. 34. Bohanec M, Messean A, Scatasta S, Angevin F, Griffiths B, et al. (2008) A qualitative multi-attribute model for economic and ecological assessment of genetically modified crops. Ecological Modelling 215: 247–261. doi: 10.1016/j.ecolmodel.2008.02.016 View Article PubMed/NCBI Google Scholar
35. Pazek K, Rozman C, Irgolic A, Turk J (2012) Multicriteria decision model for evaluating less-favoured areas for agricultural production. 47th Croatian and 7th International Symposium on Agriculture, Opatija, Croatia, 13–17 February 2012 Proceedings: 222–226.
36. Vindis P, Stajnko D, Berk P, Lakota M (2012) Evaluation of energy crops for biogas production with a combination of simulation modeling and DEX-i multicriteria method. Polish Journal of Environmental Studies

21: 763–770. doi: 10.1016/j.ecolmodel.2008.02.016 View Article PubMed/NCBI Google Scholar

37. Pazek K, Rozman C, Pavlovic V, Cerenak A, Pavlovic M (2009) The multi-criteria decision model aid for assessment of the hop cultivars (Humulus lupulus L.). In: Florijancic T, Luzaic R, editors. Zbornik Radova 44 Hrvatski i 4 Medunarodni Simpozij Agronoma, Opatija, Hrvatska, 16–20 Veljace 2009. Osijeku, Hrvatska: Poljoprivredni Fakultet Sveucilista Josipa Jurja Strossmayera u Osijeku. 360–364.
38. Pazek K, Rozman C, Cejvanovic F, Par V, Borec A, et al. (2005) Multi attribute decision model for orchard renewal - case study in Bosnia and Herzegovina. Agricultura (Slovenia) 3: 13–20. doi: 10.1016/j.ecolmodel.2008.02.016 View Article PubMed/NCBI Google Scholar
39. Tchamitchian M, Collange B, Navarrete M, Peyre G (2011) Multicriteria evaluation of the pathological resilience of soil-based protected cropping systems. In: Dorais M, editor. Acta Horticulturae. Leuven, Belgium: International Society for Horticultural Science (ISHS). 1239–1246.
40. Bohanec B (2003) Decision support. In: Mladeniæ D, Lavraè N, Bohanec M, Moyle S (Eds.). Data mining and decision support: Integration and collaboration, Kluwer Academic Publishers 23–35.
41. Delmotte S, Gary C, Ripoche A, Barbier JM, Wery J (2009) Contextualization of on farm ex-ante evaluation of the sustainability of innovative cropping systems in viticulture, using a multiple criteria assessment tool (DEXi). Wageningen University and Research Centre. 180–181.
42. Colbach N, Lucas P, Cavelier N, Cavelier A (1997) Influence of cropping system on sharp eyespot in winter wheat. Crop Protection 16: 415–422. doi: 10.1016/s0261-2194(97)00018-5 View Article PubMed/NCBI Google Scholar
43. Reinecke P, Fehrmann H (1979) Rhizoctonia cerealis van der Hoeven on cereals in the Federal Republic of Germany. Zeitschrift fur Pflanzenkrankheiten und Pflanzenschutz 86: 190–204. doi: 10.1016/s0261-2194(97)00018-5 View Article PubMed/NCBI Google Scholar
44. Cavelier N, Lucas P, Boulch G (1985) Evolution of the parasite complex Rhizoctonia cerealis Van der Hoeven and Pseudocercosporella herpotrichoides (Fron) Deighton, fungi attacking the stem base of cereals. Agronomie 5: 693–700. doi: 10.1051/agro:19850804 View Article PubMed/NCBI Google Scholar
45. Kolmer JA (2005) Tracking wheat rust on a continental scale. Current Opinion in Plant Biology 8: 441–449. doi: 10.1016/j.pbi.2005.05.001 View Article PubMed/NCBI Google Scholar

46. Bolton MD, Kolmer JA, Garvin DF (2008) Wheat leaf rust caused by Puccinia triticina. Molecular Plant Pathology 9: 563–575. doi: 10.1111/j.1364-3703.2008.00487.xView Article PubMed/NCBI Google Scholar
47. Lançon J, Wery J, Rapidel B, Angokaye M, Gerardeaux E, et al. (2007) An improved methodology for integrated crop management systems. Agronomy for Sustainable Development 27: 101–110. doi: 10.1051/agro:19850804 View Article PubMed/NCBI Google Scholar
48. Ould-Sidi MM, Lescourret F (2011) Model-based design of integrated production systems: a review. Agronomy for Sustainable Development 31: 571–588. doi: 10.1051/agro:19850804 View Article PubMed/NCBI Google Scholar
49. Pelzer E, Fortino G, Bockstaller C, Angevin F, Lamine C, et al. (2012) Assessing innovative cropping systems with DEXiPM, a qualitative multi-criteria assessment tool derived from DEXi. Ecological Indicators 18: 171–182. doi: 10.1016/j.ecolind.2011.11.019 View Article PubMed/NCBI Google Scholar
50. Sundman KE (1912) Mémoire sur le problème des trois corps, Acta Mathematica. 36: 105–179. doi: 10.1016/j.ecolind.2011.11.019 View Article PubMed/NCBI Google Scholar
51. Li BL (2010) Editorial. Ecological Complexity 7: 421–422. doi: 10.1016/j.ecolind.2011.11.019 View Article PubMed/NCBI Google Scholar
52. Madden LV (2006) Botanical epidemiology: some key advances and its continuing role in disease management. European Journal of Plant Pathology 115: 3–23. doi: 10.1016/j.ecolind.2011.11.019 View Article PubMed/NCBI Google Scholar
53. Carpani M, Bergez JE, Monod H (2012) Sensitivity analysis of a hierarchical qualitative model for sustainability assessment of cropping systems. Journal Environmental Modelling & Software 27–28: 15–22. doi: 10.1016/j.ecolind.2011.11.019 View Article PubMed/NCBI Google Scholar
54. Wallach D (2011) Crop model calibration: a statistical perspective. Agronomy Journal 103: 1144–1151. doi: 10.2134/agronj2010.0432 View Article PubMed/NCBI Google Scholar
55. Agresti A (2010) Matched-pairs data with ordered categories. In: Analysis of ordinal categorical data. Second edition. Wiley& Sons, Inc. Hoboken, New-Jersey, USA. 225–261.
56. Žnidaršič M, Bohanec M, Zupan B (2006) proDEX – A DSS tool for environmental decision-making. Environmental Modelling & Software 21: 1514–1516. doi: 10.2134/agronj2010.0432

Chapter 6

ENERGY POTENTIAL AND GREENHOUSE GAS EMISSIONS FROM BIOENERGY CROPPING SYSTEMS ON MARGINALLY PRODUCTIVE CROPLAND

[1]Marty R. Schmer mail, [2]Kenneth P. Vogel, [1]Gary E. Varvel, [3]Ronald F. Follett, [2]Robert B. Mitchell, [1]Virginia L. Jin

[1]Agroecosystem Management Research Unit, United States Department of Agriculture-Agricultural Research Service (USDA-ARS), Lincoln, Nebraska, United States of America.

[2]Grain, Forage and Bioenergy Research Unit, United States Department of Agriculture-Agricultural Research Service (USDA-ARS), Lincoln, Nebraska, United States of America.

[3]Soil-Plant Nutrient Research Unit, United States Department of Agriculture-Agricultural Research Service (USDA-ARS), Ft. Collins, Colorado, United States of America.

ABSTRACT

Low-carbon biofuel sources are being developed and evaluated in the United States and Europe to partially offset petroleum transport fuels. Current and potential biofuel production systems were evaluated from a long-term continuous no-tillage corn (Zea mays L.) and switchgrass (Panicum virgatum L.) field trial under differing harvest strategies and nitrogen (N) fertilizer intensities to determine overall environmental sustainability. Corn and switchgrass grown for bioenergy resulted in near-term net greenhouse gas (GHG) reductions of −29 to −396 grams of CO_2 equivalent emissions per megajoule of ethanol per year as a result of direct soil carbon sequestration and from the adoption of integrated biofuel conversion pathways. Management practices in switchgrass and corn resulted in large variation in petroleum offset potential. Switchgrass, using best management practices produced 3919±117 liters of ethanol per hectare and had 74±2.2 gigajoules of petroleum offsets per hectare which was similar to intensified corn systems (grain and 50% residue harvest under optimal N rates). Co-locating and integrating cellulosic biorefineries with existing dry mill corn grain ethanol facilities improved net energy yields (GJ ha^{-1}) of corn grain ethanol by >70%. A multi-feedstock, landscape approach coupled with an integrated biorefinery would be a viable option to meet growing renewable transportation fuel demands while improving the energy efficiency of first generation biofuels.

INTRODUCTION

Reduction in greenhouse gas (GHG) emissions from transportation fuels can result in near- and long-term climate benefits [1]. Biofuels are seen as a near-term solution to reduce GHG emissions, reduce U.S. petroleum import requirements, and diversify rural economies. Depending on feedstock source and management practices, greater reliance on biofuels may improve or worsen long-term sustainability of arable land. U.S. farmers have increased corn (Zea mays L.) production to meet growing biofuel demand through land expansion, improved management and genetics, increased corn plantings, or by increased continuous corn monocultures [2]–[4]. Productive cropland is finite, and corn expansion on marginally-productive cropland may lead to increased land degradation, including losses in biodiversity and other desirable ecosystem functions [4]–[6]. We define marginal cropland as fields whose crop yields are 25% below the regional average. The use of improved corn hybrids and management practices have increased U.S. grain yields by 50% since the early 1980's [7] with an equivalent increase in non-grain biomass or stover yields. Corn stover availability and expected low feedstock costs make it a likely source for cellulosic biofuel. However, excessive corn stover removal can lead to increased soil erosion and decreased soil organic carbon (SOC) [8] which can negatively affect future grain yields and sustainability. Biofuels from cellulosic feedstocks (e.g. corn stover, dedicated perennial energy grasses) are expected to have lower GHG emissions than conventional gasoline or corn grain ethanol [9]–[13]. Furthermore, dedicated perennial bioenergy crop systems such as switchgrass (Panicum virgatum L.) have the ability to significantly increase SOC [14]–[16] while providing substantial biomass quantities for conversion into biofuels under proper management [17], [18].

Long-term evaluations of feedstock production systems and management practices are needed to validate current and projected GHG emissions and energy efficiencies from the transportation sector. In a replicated, multi-year field study located 50 km west of Omaha, NE, we evaluated the potential to produce ethanol on marginal cropland from continuously-grown no-tillage corn with or without corn residue removal (50% stover removal) and from switchgrass harvested at flowering (August) versus a post-killing frost harvest. Our objectives were to compare the effects of long-term management practices including harvest strategies and N fertilizer input intensity on continuous corn grain and switchgrass to determine ethanol production, potential petroleum offsets, and net energy yields. We also present measured SOC changes (0 to 1.5 m) over a nine year period from our biofuel cropping systems to determine how direct SOC changes impact net GHG emissions from biofuels. Furthermore, we evaluate the potential efficiency advantages of co-locating and integrating cellulosic conversion capacity with existing dry mill corn grain ethanol plants.

MATERIALS AND METHODS

This study is located on the University of Nebraska Agricultural Research and Development Center, Ithaca, Nebraska, USA on a marginal cropland field with Yutan silty clay loam (fine-silty, mixed, superactive, mesic Mollic Hapludalf) and a Tomek silt loam (fine, smectitic, mesic Pachic Argiudoll) soil. Switchgrass plots were established in 1998 and continuous corn plots were initiated in 1999. The study is a randomized complete block design (replications = 3) with split-split plot treatments. Main treatments are two cultivars of switchgrass, 'Trailblazer' and 'Cave-in-Rock', and a glyphosate tolerant corn hybrid. Main treatment plots are 0.3 ha which enables the use of commercial farm equipment. Switchgrass is managed as a bioenergy crop, and corn is managed under no-tillage conditions (no-till farming since 1999). Split-plot treatments are nitrogen (N) fertilizer levels and split-split plots are harvest treatments. Annual N fertilizer rates (2000–2007) were 0 kg N ha^{-1}, 60 kg N ha^{-1}, 120 kg N ha^{-1}, and 180 kg N ha^{-1} as NH_4NO_3, broadcast on the plots at the start of the growing season. The 0 kg N ha^{-1}, 60 kg N ha^{-1}, 120 kg N ha^{-1} fertilizer rates were used on switchgrass [19] while the 60 kg N ha^{-1}, 120 kg N ha^{-1}, and 180 kg N ha^{-1} fertilizer rates were used for corn. Switchgrass harvest treatments were initiated in 2000 and consist of a one-cut harvest either in early August or after a killing frost. Corn stover treatments were initiated in 2000 and are either no stover harvest or stover removal, where the amount of stover removed approximates 50% of the aboveground biomass after corn grain is harvested.

Baseline soil samples were taken in 1998 at the center of each subplot and re-sampled in 2007 at increments of 0–5, 5–10, 10–30, 30–60, 60–90, 90–120, and 120–150 cm depths [15]. Average changes in total SOC (0–1.5 m) from 1998–2007 were used to estimate direct soil C changes. Further management practices and detailed soil property values from this study have been previously reported [15], [20]. Summary of petroleum offsets (GJ ha^{-1}), ethanol production (L ha^{-1}), greenhouse gas (GHG) emissions (g CO_2e MJ^{-1}), net GHG emissions (Mg CO2e ha^{-1}), and GHG reductions (%) for corn grain, corn grain with stover removal, and switchgrass are presented in Table S1 in File S1.

STATISTICAL ANALYSES

Yield data analyzed were from 2000 to 2007, where 2000 was the initiation of harvest treatments for continuous corn and switchgrass and 2007 was the last year that SOC was measured for this study. Data from switchgrass cultivars were pooled together based on their similar aboveground biomass yields over years and similar changes in SOC [15]. Data were analyzed using a linear mixed model approach with replications considered a random effect. Mean separation tests were conducted using the Tukey-Kramer method. Significance was set at $P \leq 0.05$.

LIFE-CYCLE ASSESSMENT

For energy requirements in the production, conversion, and distribution of corn grain ethanol and cellulosic ethanol, values from the Greenhouse Gases, Regulated Emissions, and Energy Use in Transportation (GREET v. 1.8) [21], Energy and Resources Group Biofuel Analysis Meta-Model (EBAMM) [22], and Biofuel Energy Systems Simulator (BESS) [23] life cycle assessment models were used as well as previous agricultural energy estimates for switchgrass [12]. Energy use in the agricultural phase consisted of agricultural inputs (seed, herbicides, fertilizers, packaging), machinery energy use requirements, material transport, and diesel requirements used in this study. Stover energy requirements from the production phase were from the diesel requirements to bale, load, and stack corn stover and the embodied energy of the farm machinery used. A proportion of the N fertilizer and herbicide requirements were allocated to the amount of stover harvested.

Multiple biorefinery configurations are presented to evaluate different conversion scenarios and how this affects GHG emissions, petroleum offset credits, and net energy yield (NEY) values. Biorefinery scenarios evaluated in this study are: (i) a natural gas (NG) dry mill corn grain ethanol plant with dry distillers grain (DDGS) as a co-product for the corn grain-only harvests [23]–[25], (ii) a co-located dry mill corn grain and cellulosic ethanol plant with combined heat and power (CHP) and DDGS co-product, where corn stover is primarily used to displace dry mill ethanol plant natural gas requirements [25], [26], (iii) and a standalone cellulosic (switchgrass or corn stover) ethanol plant (sequential hydrolysis and fermentation) with CHP capability and electricity export [22], [27]–[29]. Chemical and enzyme production costs and related GHG emissions for corn grain and cellulosic conversion to ethanol were also incorporated [28]. Ethanol recovery for corn grain was estimated to be 0.419 L kg^{-1} [23]. Ethanol recovery for corn stover and switchgrass were based on cell wall composition from harvested biomass samples. Ground aboveground switchgrass samples were scanned using a near-infrared spectrometer to predict cell wall and soluble carbohydrate biomass composition [30]. Ground corn stover samples were analyzed using a near-infrared spectrometer-based calibration equation developed by the National Renewable

Energy Laboratory to predict corn stover cell wall composition [31]. Switchgrass and corn stover cell wall conversion to ethanol was based on composition components of glucan, xylose and arabinose [30], [31]. Glucan to ethanol conversion was assumed to be 85.5%, and xylose and arabinose was estimated to have 85% ethanol recovery efficiency [29]. Estimated ethanol recovery for corn stover was 327 L Mg−1 which was similar to other findings [29]. For switchgrass, ethanol recovery based on glucan, xylose, and arabinose concentrations was estimated to be 311 L Mg^{-1} and 344 L Mg^{-1} for an August harvest and a post-frost harvest, respectively.

Ethanol plant size capacity was estimated to be 189 million L yr−1 for the corn grain-only and cellulosic-only scenarios. For the co-located facility, total plant size was assumed to be 378 million L yr−1 capacity. Fossil fuel energy requirements for the conventional corn grain ethanol plant is assumed to be 7.69 MJ L−1 for natural gas to power the plant and to dry DGS, 0.59 MJ L−1 for corn grain transportation from farm to ethanol plant, 0.67 MJ L−1 for electricity purposes, 0.13 MJ L−1 to capital depreciation costs, and 0.58 MJ L−1 for wastewater processing and effluent restoration [10], [22]. Fossil fuel requirements for the corn grain/cellulosic ethanol plant are feedstock transportation 0.63 MJ L−1 for corn stover, 0.59 MJ L−1 for corn grain transportation from farm to ethanol plant, 0.44 MJ L−1 to capital depreciation costs, and 0.58 MJ L−1 for wastewater treatment and processing (Table S2 in File S1). Cellulosic ethanol plant fossil fuel requirements are 0.63 MJ L−1 for switchgrass transportation from field to ethanol plant, 0.06 MJ L−1 diesel requirements for biomass transport within the ethanol plant grounds, 0.44 MJ L−1 to capital depreciation costs, and 0.58 MJ L−1 for wastewater processing, effluent restoration, and recovery (Table S2 in File S1).

For the co-located corn grain and cellulosic facility, we assumed (i) power and electrical utilities were shared [26]; (ii) power requirements were supplied mainly from the lignin portion of stover with combined ethanol purification from the starch and cellulosic ethanol conversion pathways [26]; and (iii) extra stover biomass would be required in addition to the lignin to meet steam requirements. A co-location facility would require additional unprocessed bales to be used in addition to lignin which lowered the amount of ethanol being generated from stover at a co-located facility compared to a standalone cellulosic facility that uses stover as their primary feedstock (Table S1 in File S1). Electricity would be imported from the grid in this scenario and DDGS exported as the only co-product. Recent analysis [29] of converting cellulose to ethanol has estimated a higher internal electrical demand than previously assumed [26]; suggesting electricity export under this configuration would be unlikely. The value of DDGS as animal feed would likely preclude its use in meeting power requirement in a co-located facility. We based our total biomass energy requirement on the lignin concentration in stover and the expected biomass energy use requirements to power a co-located ethanol plant [25]. Estimated biomass requirements were 11 MJ L−1 ethanol and embodied energy value of 16.5 MJ kg−1 (low heating value) for stover biomass.

Net energy yield (NEY) values (renewable output energy - fossil fuel input energy) were calculated for each feedstock and conversion scenario. Output energy was calculated from ethanol output plus co-product credits. Co-product credit for DDGS is 4.13 MJ L−1 for the corn grain-only ethanol plant and the co-located corn grain/cellulosic ethanol plant [32]. Electricity co-product credit for standalone cellulosic ethanol was estimated to be 1.68 MJ L−1 [29]. Petroleum offsets (GJ ha−1) were calculated in a similar fashion as NEY with total ethanol production (MJ ha−1) along with petroleum displacement from co-products minus petroleum inputs consumed in the production, conversion, and distribution phase (Tables S1 and S3 in File S1). Petroleum offsets were calculated as the difference between ethanol output and petroleum inputs from the agricultural, conversion, and distribution phase (Table S1 in File S1). Petroleum requirements for each cropping system were calculated from input requirements from this study and derived values from the EBAMM model [22]. For input requirements without defined petroleum usage, we used the default parameter in EBAMM that estimates U.S. average petroleum consumption at 40% for input source. Petroleum offset credits associated with corn grain ethanol co-products were estimated to be 0.71 MJ L−1 while credits for corn stover and switchgrass cellulosic ethanol co-products (standalone facility) were 0.12 MJ L−1 (Table S3 in File S1). Petroleum offset credits were calculated from GREET (v 1.8).

GREENHOUSE GAS EMISSIONS

Greenhouse gas offsets associated with the production of corn grain and cellulosic ethanol were modeled from the EBAMM and BESS models [22], [23]. Agricultural GHG emissions were based on fuel use, fertilizer use, herbicide use, farm machinery requirements, and changes in SOC. Direct land use change by treatment plot can either be a GHG source or a GHG sink depending on SOC changes from this study [15]. Co-product GHG credits for DDGS or electricity export were derived from the BESS [23] and GREET (v. 1.8) models [21]. Co-product GHG credits for DDGS was −347 g CO2e L−1 ethanol and −304 g CO2e L−1 ethanol for cellulosic electricity export (Table S4 in File S1). Indirect land use changes for corn grain ethanol or switchgrass were not estimated in this analysis. GHG offsets were calculated on both an energy and areal basis (Table S1 in File S1).

Greenhouse gas emissions from N fertilizer were evaluated from the embodied energy requirements and subsequent nitrous oxide (N_2O) emissions (Table S4 in File S1). Direct and indirect nitrous oxide emissions were calculated in this study using Tier 1 Intergovernmental Panel on Climate Change calculations. Greenhouse gas emission values for the agricultural phase are included in Table S4 in File S1 and for the conversion and distribution phase in Table S5 in File S1. For the agricultural phase, total GHG emissions were calculated from the production of fertilizers, herbicides, diesel requirements, drying costs for corn grain, and the embodied energy in farm machinery minus direct soil C changes occurring for the study period (Table S4 in File S1). GHG emissions were reported on an energy basis, areal basis, and the difference between ethanol and conventional gasoline (Table S1 in File S1). For net GHG emissions (Mg CO2e ha−1), calculations were based on GHG intensity values (g CO2e MJ−1) multiplied by biofuel production (MJ ha−1) for each cropping system. GHG reductions (Table S1 in File S1) were calculated as the percent difference from conventional gasoline as reported by the California Air Resource Board (99.1 g CO2 MJ−1) [33].

RESULTS AND DISCUSSION

Harvest and N fertilizer management treatments affected grain and biomass yields in both crops over eight growing seasons (Fig. 1A).

Switchgrass harvested after a killing frost had 27% to 60% greater biomass yields compared with an August harvest under similar fertilization rates. Highest harvested biomass yields (mean = 11.5 Mg ha−1 yr−1) were from fertilized (120 kg N ha−1) switchgrass harvested after a killing frost while continuous corn showed similar grain and stover yields [factorial analysis of variance (ANOVA), P = 0.72] under the highest N fertilizer levels (180 kg N ha−1) (Fig. 1A).

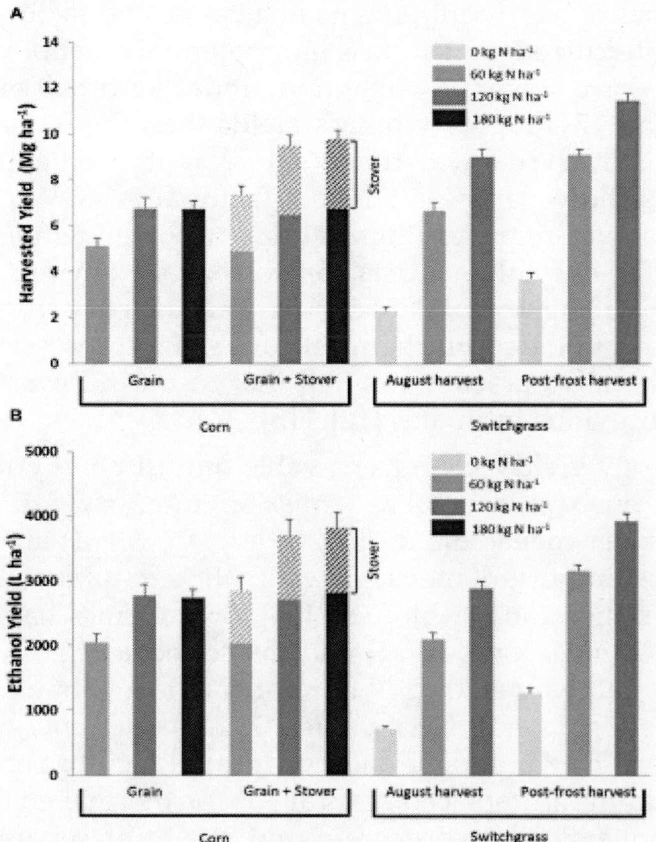

Figure 1 Harvested mean annual yield ± standard error (A) and ethanol energy ± SEM (B) for no-till continuous corn (grain-only harvest or grain and stover harvest) and switchgrass (August harvest or Post-frost harvest) under variable nitrogen rates on marginally-productive rainfed cropland for 2000–2007 (n = 3 replicate corn system plots and 6 replicate switchgrass plots).

doi:10.1371/journal.pone.0089501.g001

Potential ethanol yields varied from 2050 to 2774 L ha−1 yr−1 for corn grain-only harvests while those for corn grain with stover removal ranged from 2862 to 3826 L ethanol ha−1 yr−1 (Fig. 1B). Ethanol contribution from corn stover ranged from 820 to 998 L ha−1 yr−1 when stover is converted at a standalone cellulosic plant (Fig. 1B). Separate ethanol facilities showed slightly higher potential ethanol yields (L ha−1) than at a co-located facility (Table S1 in File S1) because a larger portion of corn stover biomass was required to meet thermal power requirements at a co-located facility (SI text in File S1). Unfertilized switchgrass had potential ethanol yield values similar to corn stover. Switchgrass under optimal management practices had 17% higher biomass yields than the highest yielding corn with stover removal treatment. Potential ethanol yield for switchgrass, however, was similar (factorial ANOVA, $P>0.05$) to corn with stover removal (Fig. 1B) due to lower cellulosic ethanol recovery efficiency than exists for corn grain ethanol conversion efficiency. Switchgrass ethanol conversion efficiency from this study was based on updated biochemical conversion processes [29] using known cell wall characteristics [30] that result in lower conversion rates than previous estimates [12], [18].

Net energy yield (NEY) (renewable output energy minus fossil fuel input energy) and GHG emission intensity (grams of CO_2 equivalents per megajoule of fuel, or g CO2e MJ−1) are considered the two most important metrics in estimating fossil fuel replacement and GHG mitigation for biofuels [34]. Switchgrass harvested after a killing frost (120 kg N ha−1) and the co-located grain and stover conversion pathway (120 kg N ha−1 and 180 kg N ha−1 treatments) had the highest overall NEY values (Fig. 2). Net energy yields for continuous corn were higher at a co-located facility because stover biomass and lignin replaced natural gas for thermal energy (Fig. 2). Ethanol conversion of corn grain and stover at separate facilities was intermediate in NEY while traditional corn grain-only natural gas (NG) dry mill ethanol plants had the lowest NEY values for the continuous corn systems. Delaying switchgrass harvest from late summer to after a killing frost resulted in significant improvement in NEY and potential ethanol output under similar N rates. Unfertilized switchgrass had similar NEY values compared with corn grain processed at a NG dry mill ethanol plant (factorial ANOVA, $P = 0.12$) while fertilized switchgrass harvested after a killing frost had higher NEY values (factorial ANOVA, P<0.0001) than NG dry mill corn grain ethanol plants (Fig. 2).

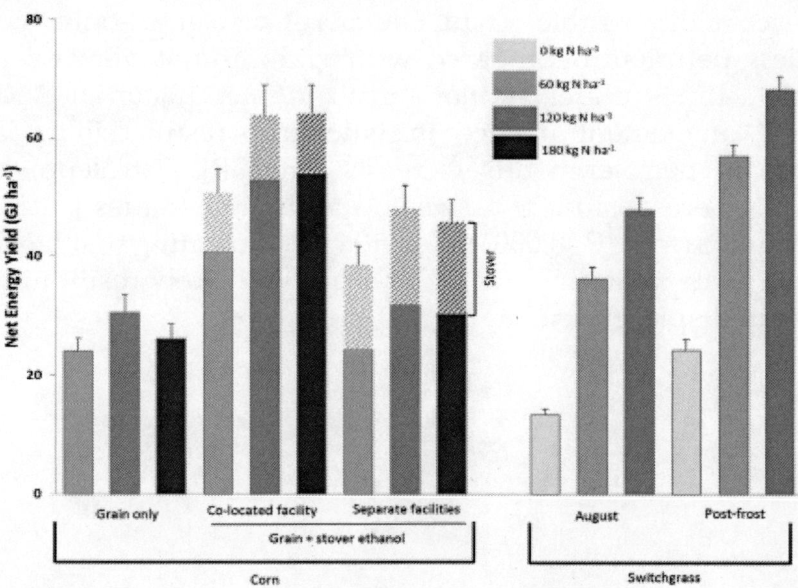

Figure 2 Net energy yield ± standard error for no-till continuous corn (grain-only or grain and stover harvest) and switchgrass (August harvest or post-frost harvest) under variable nitrogen rates on marginally-productive cropland (n = 3 replicate corn system plots and 6 replicate switchgrass plots).

Conversion processes evaluated include corn grain-only harvest at a natural gas (NG) dry mill, corn grain with stover harvest at a co-located facility (lignin portion of stover used as primary energy source for grain and cellulose conversion), corn grain with stover harvest at separate ethanol facilities (NG dry mill and a cellulosic ethanol plant), and switchgrass (cellulosic ethanol plant).

doi:10.1371/journal.pone.0089501.g002

Both the continuous corn and switchgrass systems showed significant petroleum offset (ethanol output minus petroleum inputs) capability, with the intensified bioenergy cropping systems having the highest petroleum offsets (Fig. 3). Petroleum use varied by cropping system in the agricultural phase with continuous corn systems having higher overall petroleum requirements than switchgrass. Petroleum requirements (mainly diesel fuel) to harvest corn stover are small relative to corn grain harvest as a result of low harvested stover yields. Lowest petroleum offsets for continuous corn systems were from stover harvests at a separate dedicated cellulosic facility (Table S1 in File S1). Corn grain-only harvests offset less petroleum compared with grain and stover at separate ethanol facilities under similar fertilizer rates (factorial ANOVA, P<0.01). Management practices in switchgrass resulted in the largest variation in petroleum offset credits (Fig. 3B). Petroleum offsets (GJ ha−1) were positively associated with NEY values [−1.81+0.84 (Petroleum offset); (P<0.0001); (R2 = 0.76)], indicating that bioenergy cropping systems with large NEY values will likely result in higher petroleum displacement.

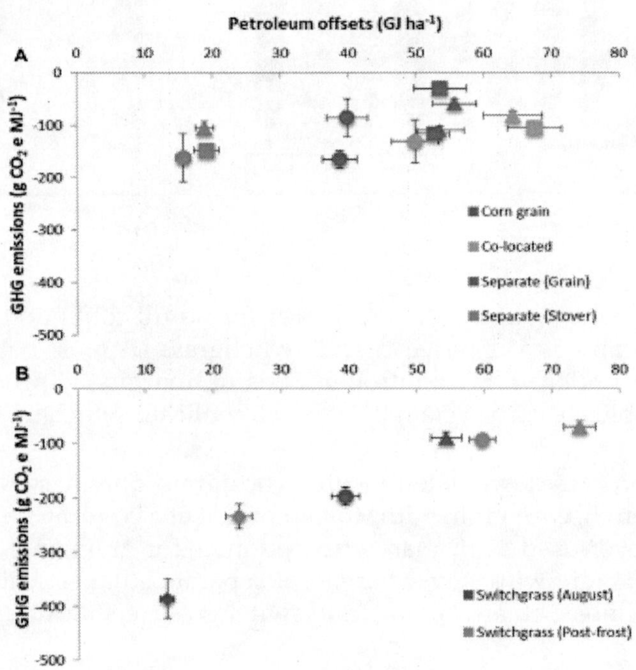

Figure 3 Petroleum offsets compared with GHG emissions (g CO_2e MJ^{-1} ethanol) for continuous corn and switchgrass grown on marginally-productive cropland (n = 3 replicate corn system plots and 6 replicate switchgrass plots).

(A) Continuous corn values represent harvest method (stover harvested or retained) and ethanol conversion pathway (co-located facility or at a separate ethanol facilities). (B) Switchgrass values are based on harvest date and N fertilizer rate. Fertilizer rates are 0 kg N ha−1 (♦), 60 kg N ha−1 (•), 120 kg N ha−1 (□), and 180 kg N ha−1 (▪). Error bars indicate standard errors of the mean.

doi:10.1371/journal.pone.0089501.g003

All bioenergy cropping systems evaluated in our study had SOC sequestration rates exceeding 7.3 Mg CO2 yr−1 (Table S4 in File S1), with over 50% of SOC sequestration occurring below the 0.3 m soil depth [15]. Soil organic C increased even with corn stover removal, indicating that removal rates were sustainable in terms of SOC and grain yield for this time period. No-tillage continuous corn systems have lower stover retention requirements to maintain SOC than continuous corn with tillage or corn-soybean (Glycine max (L.) Merr.) rotations [8]. Consequently, all conversion pathways had negative GHG emission values as a result of SOC sequestration offsetting GHG emissions from the production, harvest, conversion and distribution phases for corn grain ethanol and cellulosic ethanol. For switchgrass, SOC storage values were similar to other findings within the same ecoregion [16] and a long-term Conservation Reserve Program grassland [35]. Measured SOC storage from the continuous corn systems (Table S4 in File S1) were significantly higher than modeled SOC storage estimates from this region [36]. Corn grain grown with low N rates (60 kg ha−1) had GHG intensity values similar to continuous corn under optimum N rates (120 kg ha−1) but resulted in lower ethanol yields and lower petroleum offset potential (Fig. 3A). Lowest GHG emission intensity values on an energy basis (g CO2e MJ−1) were from unfertilized switchgrass (Table S1 in File S1) due to lower ethanol yields, lower agricultural energy emissions, and similar SOC storage compared with the other biofuel cropping systems. For switchgrass, management practices that resulted in the lowest GHG emission on an energy basis resulted in the lowest petroleum offset potential (Fig. 3B). Direct N_2O emissions (Table S4 in File S1) were estimated using Intergovernmental Panel on Climate Change methodology and are in agreement with study site N2O flux measurements from a later time series which indicated N rate as the major contributor to N_2O emissions [37]. When evaluating GHG emissions on a per unit area basis (g CO_2e ha^{-1}), unfertilized switchgrass and corn grain-only systems showed similar results with the more intensified cropping systems (Table S1 in File S1).

Both switchgrass and continuous corn with stover removal produced similar ethanol potential, NEY values, petroleum offsets, and GHG emissions but overall values and metric efficiencies were dependent on management practices and downstream conversion scenarios. Dedicated perennial grass systems used for bioenergy will need to have similar or greater yield potential than existing annual crops for widespread adoption to meet renewable energy demands and provide similar economic returns to producers. We have previously shown that switchgrass ethanol yields were comparable with regional corn grain ethanol yields [12]. Here we demonstrate that when switchgrass is optimally managed, ethanol potential is similar to a continuous corn cropping system with stover removal and exceeds ethanol yield for corn grain-only systems on marginally-productive cropland. Furthermore, breeding improvements for bioenergy specific switchgrass cultivars have shown higher yield potential than cultivars evaluated here [38].

Coupling sustainable agricultural residue harvests with dedicated energy crops improves land-use efficiency and reduces biomass constraints for a mature cellulosic biofuel industry. Recent analysis has shown that sufficient land exists in the U.S. Corn Belt to support a cellulosic ethanol industry without impacting productive cropland [18], [39], [40]. The effect of dedicated energy crops and corn grain on indirect land use change varies significantly based on the assumptions and models used [13], [41], [42] but bioenergy crops grown on marginally-productive cropland will have less impact on indirect land use change than bioenergy crops grown on more productive cropland. Likewise, model assumptions underlying direct SOC sequestration will impact system evaluations of GHG emissions and mitigation. Measured SOC sequestration values presented here were based on production years evaluated and were not extrapolated beyond this time-frame. Extrapolating SOC values from this time-frame to a 30-yr time horizon or 100-yr time horizon is still larger than current life cycle assessment assumptions on SOC sequestration potential of switchgrass or no-till corn [12], [42], [43]. This highlights the importance of accounting for direct SOC changes at depth to accurately estimate GHG emissions for biofuels under both marginal and productive cropland. Further long term evaluation of management practices (e.g. tillage, stover removal) on

SOC sequestration potential for corn grain systems under irrigated conditions on productive cropland is warranted [44].

A multi-feedstock, landscape approach minimizes economic and environmental risks in meeting feedstock demands for cellulosic ethanol production by providing sufficient feedstock availability while maintaining ecosystem services. A co-located cellulosic biorefinery is expected to have economic advantages by reducing capital costs requirements for cellulosic conversion and through sharing of infrastructure costs. In this study, we used corn stover as the feedstock for the co-located cellulosic biorefinery but the benefits will apply to other cellulosic feedstocks. A co-located facility can increase NEY values by decreasing natural gas use for thermal energy, but current and forecasted U.S. natural gas prices [45] may affect large scale adoption of co-location unless there are incentives for displacing fossil energy in existing NG dry mill ethanol plants [46]. Integrating cellulosic refining capacity with existing corn grain ethanol plants can improve the sustainability of first generation biofuels and enable the implementation of cellulosic biofuels into the U.S. transportation sector.

ACKNOWLEDGMENTS

Authors would also like to acknowledge the contribution of Edward J. Wolfrum and his staff at the National Renewable Energy Laboratory (Golden, CO) in conducting the corn stover analysis. Mention of trade names or commercial products in this publication does not imply recommendation or endorsement by the U.S. Department of Agriculture. USDA is an equal opportunity provider and employer.

AUTHOR CONTRIBUTIONS

Conceived and designed the experiments: KV GV RF. Performed the experiments: KV GV RF RM MS VJ. Analyzed the data: MS. Contributed reagents/materials/analysis tools: MS KV RM RF. Wrote the paper: MS KV GV RF RM VJ.

REFERENCES

1. Unger N, Bond TC, Wang JS, Koch DM, Menon S, et al. (2010) Attribution of climate forcing to economic sectors. Proc Natl Acad Sci USA 107(8): 3382–3387. doi: 10.1073/pnas.0906548107 View Article PubMed/NCBI Google Scholar
2. Wallander S, Claassen R, Nickerson C (2011) The ethanol decade: An expansion of U.S. corn production, 2000–09. USDA-ERS Economic Information Bulletin No. (EIB-79).
3. Claassen R, Carriazo F, Cooper J, Hellerstein D, Ueda K (2011) Grassland to cropland conversion in the northern plains: The role of crop insurance, commodity, and disaster programs Economic Research Report No. (ERR-120).
4. Wright CK, Wimberly MC (2013) Recent land use change in the western corn belt threatens grasslands and wetlands. Proc Natl Acad Sci USA 110(10): 4134–4139. doi: 10.1073/pnas.1215404110 View Article PubMed/NCBI Google Scholar
5. Wiens J, Fargione J, Hill J (2011) Biofuels and biodiversity. Ecol Appl 21(4): 1085–1095. doi: 10.1890/09-0673.1 View Article PubMed/NCBI Google Scholar
6. Donner SD, Kucharik CJ (2008) Corn-based ethanol production compromises goal of reducing nitrogen export by the Mississippi river. Proc Natl Acad Sci USA 105(11): 4513–4518. doi: 10.1073/pnas.0708300105 View Article PubMed/NCBI Google Scholar
7. USDA-NASS. 2013 U.S. Corn Yields. Available: http://www.nass.usda.gov/Charts_and_Maps/Field_Crops/cornyld.asp.
8. Wilhelm WW, Johnson JMF, Karlen DL, Lightle DT (2007) Corn stover to sustain soil organic carbon further constrains biomass supply. Agron J 99(6): 1665–1667. doi: 10.2134/agronj2007.0150 View Article PubMed/NCBI Google Scholar
9. Adler PR, Del Grosso SJ, Parton WJ (2007) Life-cycle assessment of net greenhouse-gas flux for bioenergy cropping systems. Ecol Appl 17(3): 675–691. doi: 10.1890/05-2018 View Article PubMed/NCBI Google Scholar
10. Farrell AE, Plevin RJ, Turner BT, Jones AD, O'Hare M, et al. (2006) Ethanol can contribute to energy and environmental goals. Science 312(5781): 506–508. doi: 10.1126/science.1121416 View Article PubMed/NCBI Google Scholar
11. Sheehan J, Aden A, Paustian K, Killian K, Brenner J, et al. (2003) Energy and environmental aspects of using corn stover for fuel ethanol. J

Indust Ecol 7(3–4): 117–146. doi: 10.1162/108819803323059433 View Article PubMed/NCBI Google Scholar

12. Schmer MR, Vogel KP, Mitchell RB, Perrin RK (2008) Net energy of cellulosic ethanol from switchgrass. Proc Natl Acad Sci USA 105(2): 464–469. doi: 10.1073/pnas.0704767105 View Article PubMed/NCBI Google Scholar
13. Wang M, Han J, Dunn JB, Cai H, Elgowainy A (2012) Well-to-wheels energy use and greenhouse gas emissions of ethanol from corn, sugarcane and cellulosic biomass for US use. Environ Res Letters 7(4): 045905. doi: 10.1088/1748-9326/7/4/045905 View Article PubMed/NCBI Google Scholar
14. Frank AB, Berdahl JD, Hanson JD, Liebig MA, Johnson HA (2004) Biomass and carbon partitioning in switchgrass. Crop Sci 44: 1391–1396. doi: 10.2135/cropsci2004.1391 View Article PubMed/NCBI Google Scholar
15. Follett RF, Vogel KP, Varvel GE, Mitchell RB, Kimble J (2012) Soil carbon sequestration by switchgrass and no-till maize grown for bioenergy. BioEnerg Res 5: 866–875. doi: 10.1007/s12155-012-9198-y

Chapter 7

A GENERIC MODEL TO SIMULATE AIR-BORNE DISEASES AS A FUNCTION OF CROP ARCHITECTURE

[1]Pierre Casadebaig, [1]Gauthier Quesnel, [2]Michel Langlais, [1]Robert Faivre

[1]Unité de Biométrie et Intelligence Artificielle UPR 875, Institut National de la Recherche Agronomique, Centre de Recherches de Toulouse, Castanet-Tolosan, France

[2]Institut de Mathematiques de Bordeaux, Université de Bordeaux, Bordeaux, France

ABSTRACT

In a context of pesticide use reduction, alternatives to chemical-based crop protection strategies are needed to control diseases. Crop and plant architectures can be viewed as levers to control disease outbreaks by affecting microclimate within the canopy or pathogen transmission between plants. Modeling and simulation is a key approach to help analyze the behaviour of such systems where direct observations are difficult and tedious. Modeling permits the joining of concepts from ecophysiology and epidemiology to define structures and functions generic enough to describe a wide range of epidemiological dynamics. Additionally, this conception should minimize computing time by both limiting the complexity and setting an efficient software implementation. In this paper, our aim was to present a model that suited these constraints so it could first be used as a research and teaching tool to promote discussions about epidemic management in cropping systems. The system was modelled as a combination of individual hosts (population of plants or organs) and infectious agents (pathogens) whose contacts are restricted through a network of connections. The system dynamics were described at an individual scale. Additional attention was given to the identification of generic properties of host-pathogen systems to widen the model's applicability domain. Two specific pathosystems with contrasted crop architectures were considered: ascochyta blight on pea (homogeneously layered canopy) and potato late blight (lattice of individualized plants). The model behavior was assessed by simulation and sensitivity analysis and these results were discussed against the model ability to discriminate between the defined types of epidemics. Crop traits related to disease avoidance resulting in a low exposure, a slow dispersal or a de-synchronization of plant and pathogen cycles were shown to strongly impact the disease severity at the crop scale.

INTRODUCTION

In agriculture, growers have to compete with harmful organisms (animal pests, plant pathogens and weeds), collectively called pests [1]. These pests have different ways to injure plants [2] and cause crop losses (quantitative or qualitative damages), causing economic losses. Among pathogens, diseases related to fungi can severely impair plant physiology by lowering light interception or reducing photosynthetic rate.

Thus, yield increases in the last decades resulted essentially from genetic progress and massive use of pesticides. However, current practices in crop protection generate a set of issues: (1) economical, as the cost of chemical treatments lowers margins when crop prices or yields are low; (2) durability, as the use of pesticides causes strong environmental risks: some pathogen populations can adapt and gain resistance rapidly (around 4 years); (3) technical, as the integration of alternative methods of crop protection (genetic, chemical, physical, biological and cultural control) is hard to achieve in a farm and finally (4) scientific, as the joint management of plant and pathogen populations in time (crop rotations) and space (landscape structure,

[3]) is still a research question (integrated avirulence management, [4], [5]).

In a context of pesticide use reduction, alternatives to chemical-based crop protection strategies are needed to control diseases. Before the pesticide era, the canopy structure - resulting from the plant architecture and crop management - was the main natural lever to try to control disease development in cultivated fields. An update of this expertise is proposed through a research project focused on four architecturally contrasted systems and their main associated diseases: pisum-aschochyta blight, potato-late blight, vineyard-powdery mildew and yam-anthracnose systems. The first two systems were choosen as case studies for this paper. Knowledge in epidemiology tends to be very specific to the studied system. Modeling can help identify and share similarities between systems to unravel crop and pathogen interactions on a wider range of systems. Later, as a more operational step, modeling could be used to quantify how the canopy structure affects disease outbreaks and to identify architectural traits or cropping practices that lower disease severity. Provided that the model embeds genetic information (genetic basis of phenotypic parameters) it may later become a tool to design adapted genotypes [6], [7].

This scientific and operational context set several constraints on the model development. The main objective of this work was to cross the point of view of ecophysiologists and pathologists on the canopy function and structure to design a model with a generic scope, i.e. based on a single representation for different pathosystems, to explain contrasted dynamics of epidemics. Secondly, as the analysis to understand and use the model would be heavily based on simulation, the model design should minimize computing time by both limiting the complexity and setting an efficient software implementation. Our aim in this paper was to present a model that suited these constraints so it can first be used as a research and teaching tool to promote discussions about the control of diseases.

We first present the conceptual design phase (Analysis section) that defines constraints for the following phase of model design and implementation (Results section). Model evaluation is presented in the Discussion section before presenting conclusions and perspectives.

ANALYSIS

Essentially, a crop-pathogen system can be viewed as the juxtaposition of individual hosts (plants) and infectious agents (pathogens) whose contacts are restricted through dispersal network; these three elements evolving under a fluctuating climate. By adopting a shared point of view on the crop and the pathogen, this dispersal network is defined by both the crop structure and the pathogen's dispersal mode. The properties of the generic objects defining the system are here detailed and are the conceptual basis of the model. This conceptual basis merges basic concepts from epidemiology, crop modeling and graph theory. The structure and dynamics of the formal model are presented in the Results section.

Merging Concepts from Basic Epidemiology, Crop Modeling, and Graph Theory

Brief history of epidemic models in phytopathology

When observing the evolution of disease severity at the field scale, the crop leaf area is usually logistically destructed by the pathogen. As reviewed by Hau [8], two main approaches assuming different levels of complexity were investigated in the bibliography to model this observation: (1) mechanistic detailed models dealing with small-scale processes to understand the functioning of the whole system and (2) analytical models, based mostly on differential equations, with a limited number of processes and parameters [9]. The detailed approach tends to be very specific of the system (crop, pathogen species). An important feature of our approach is to build a model suitable for different systems, so we focused on the latter approach, based on simpler dynamics.

In the first models from Vanderplank [9] the evolution of disease severity ( ) can be described by the logistic growth rate as a function of time ( ) and an infection rate ( ). Such simple growth functions can be used to describe epidemics but not to explain the underlying biological processes [8]. Customized models could be built from this starting point to include biological processes such as the partitioning of the diseased individuals into different compartments (healthy, latent, infectious and removed) [10], namely SEIR models in human epidemiology. The interaction between host and pathogen growth could also be added to simpler models, as initiated by Jeger [11]. The main hypothesis underneath this model was the potential equilibrium state between host and pathogen populations. Such model design could extend to the integration of different organization scales still without using explicitly spatialized variables [12].

If further complexity is added in theoretical models, representing a generic behavior of a pathogen class (fungi) is trickier and less pertinent. Consequently, the computational study of the model is also more complicated. In analytical models, parameters such as in the logistic growth contains information about the contacts underlying the disease transmission process. In this case, random contacts are assumed in the population (random mixing network). Considering non-mobile hosts, the number of contacts that each individual has is considerably smaller than the population size and, in such circumstances, random mixing does not occur [13]. It is thus important to model the contact network that depends not only on the type of pathogen's dispersion (wind, splashing, ...) but also on the host's spatial arrangement.

Representing the canopy structure with a variable scale object

Plant epidemiology [14] and crop modeling [15], [16] are very active research topics but approaches that link these two domains are less frequent [2], [17], [18]. As the case studies for these models range from improving scientific knowledge to engineering, finding an appropriate coupling scale is difficult. For example, the effect of plant architecture on disease dispersal and severity was modeled by coupling a structural plant model (3D "virtual plant") and experimental disease inoculations [19]. This organ-scale model would not be the most appropriate basis for adding dynamics of host or pathogen growth that would fit different crops, because of the overwhelming parameters fit needed. In our case study, we will use modelling concepts from crop models at a broad scale to allow an easier setup of generic features.

We assume that an approximation of the geometry rather than a fully detailed 3D plant structure is an acceptable simplification when considering the crop at an integrative level. When dealing with processes at crop scale, such as epidemic development, the underlying complexity could partly be discarded. If complexity is ignored, the model is able to describe the epidemic (using logistic growth for example) but significant interactions between elements of the system will not be highlighted.

We have chosen to keep sub-crop scale objects (plants or organs) but discard their geometric attributes (size, shape, spatial position and orientation). The canopy was represented by an object that was simply named functional unit, the scale of this object was relative to the epidemic type. It was also more convenient to describe processes at the level where they are better known for each system. For example, in dense canopy, where epidemics tend to ascend from lower crop layers [20], individual plants are less pertinent than a representation of a mean plant divided in stem segments (phytomers). In patchy epidemics, individual plants get infected and disperse infectious agents in their local neighbourhood. These two epidemic archetypes were taken as a case study. Thus, functional units can represent either a mean plant, individual plants or a sub-part of a plant. The canopy architecture was characterized by a network of functional units. Dynamics of plant and pathogen growth were described on functional units, meaning that a homogeneous behavior of plant tissues was supposed within the unit.

Modeling Spore Dispersal by Using a Graph-theory Perspective

Early epidemiological models (logistic growth or SEIR-derived) have a long and successful history but are based on population-wide random mixing [13]. In agronomic systems, where canopy results mainly from interaction between cropping practices and the environment, there is often a strong spatial structure of the host population (rows, trellis). In practice, each individual has a finite set of contacts that they can infect. An analogy with graph theory, with hosts as graph nodes and neighbourhood of infection as edges, seems an adapted approach to deal with these limited contacts [13], [21].

In our model, a graph, namely the network of connections between functional units, was used to represent the spore dispersal process.

Setting up an experiment to determine the real network of connection is tedious for spore-transmitted infections as relationships between individuals are not observable in field. Therefore, the representation of the real network results from an a priori point of view on both crop structure and dispersal mode rather than direct observation.

RESULTS

Model Overview

The epidemiological model consists of individual-based sub-models for pathogen and leaf area dynamics linked together by a network that represents the canopy architecture. Figure 1 shows these two scales, a global one (crop) and a local one (individual hosts referred to as functional units) which interplay by using common attributes. This conception allowed individual host dynamics to be affected by global crop properties that are in turn dependent of individual attributes. Figure 1 presents a static view of the model before the beginning of the simulation. As the simulation progresses, functional units are created and linked together according to the specifications of the connection network.

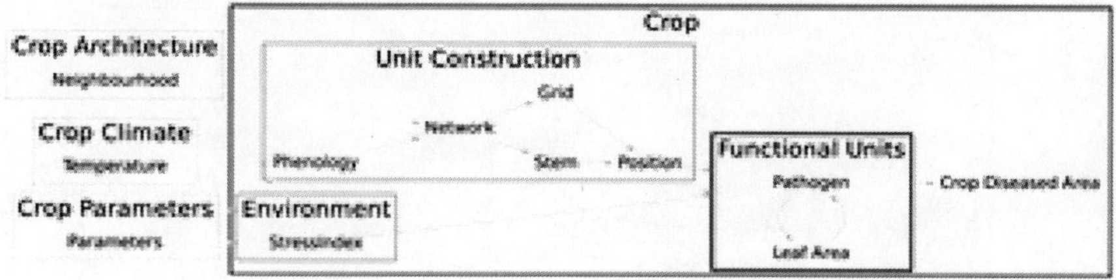

Figure 1 General model structure.

This summary view represents a hierarchical organization of sub-models which describes objects and processes of the system. Model inputs are shown in dotted frames. Scales (crop and unit) are shown in bold lines. Sub-model

Unit Construction acts as a controller for the dynamically created and individual based models Functional Units.

doi:10.1371/journal.pone.0049406.g001

The canopy structure and development processes are detailed in the Unit Construction sub-model. The Functional Units model describes the dynamics of plant (leaf area) and pathogen (leaf area injured) growth at local scale as a set of ordinary differential equations. The environmental effect on pathogen growth is computed in the Environment sub-model.

The formal model is deterministic with a discrete time-step (daily). Model's variables are summarized in table 1. System input variables are climatic measurements (mean air temperature). Input parameters are defined in table 2.

Table 1 Summary table for variables

symbol	name	unit	eq.
T_m	mean air temperature	°C	
t	thermal time	°C.d	
	unit number (parameter in "grid" method)	-	
LAI	leaf area index	m^2/m^2	
t	thermal age	°C.d	1, 2, 3, 5
Z	height	m	1
ϕ	porosity	-	6
ρ	receptivity	-	5
θ	thermal stress	-	8
δi_d	allo-deposition	m^2	7
$\tau_a \cdot I$	auto-deposition	m^2	4
H	healthy area	m^2	4
L	latent infected area	m^2	4
I	infectious area	m^2	4
R	removed area	m^2	4
E	expanded area	m^2	2
A	photosynthetically active area	m^2	
S	senescent area	m^2	3

Horizontal lines separate crop attributes from unit specific attributes. Rate-based variables directly linked to state variables were not specified in the table.
doi:10.1371/journal.pone.0049406.t001

doi:10.1371/journal.pone.0049406.t001

Table 2 Summary table for parameters

symbol	name	stem	grid	uncertainty	unit	eq.
T_b	base temperature	0	4		°C	
t_c	phyllochrons	70 : 50	NA		°C.d	
t_F	crop thermal time to flowering	1500	2000		°C.d	
Z_{max}	potential height of unit	1	1		m	1
k_z	increase of elongation rate	0.01	0.013		°C^{-1}d^{-1}	1
t_z	unit / crop thermal time for $Z_{max}/2$	300	1000		°C.d	1
E_{max}	potential area of unit	1	1		m^2	2
k_e	increase of expansion rate	0.01	0.013		°C^{-1}d^{-1}	2
t_e	unit / crop thermal time for $E_{max}/2$	300	1000		°C.d	2
k_s	increase of senescence rate	0.03	0.06		°C^{-1}d^{-1}	3
t_s	unit / crop thermal time for $E_{max}/2$	1000	2500	[70; 130] %	°C.d	3
i_p	duration of infectious period	10	10	[1;15]	d	4
l_p	duration of latency period	3	5	[1;15]	d	4
A_i	quantity of inoculation	0.01	0.01	[0.01; 0.3]	m^2	
D_i	day of disease initiation	50	50	[30; 140]	d	
c_e	scalar for between-unit infection rate	0.4	0.5	[0.05; 1]	-	4
c_w	scalar for within-unit infection rate	0.6	0.5	[0.05; 1]	-	4
t_r	shape of receptivity function	500	500	[100; 800]	°C.d	5
k_r	slope of receptivity function	0.05	0.01		°C^{-1}d^{-1}	5
ϕ_s	scalar for porosity level	0.8	0.9	[0.7; 1]	m$^{-1/2}$	6
T_{opt}	optimal temperature for infection	23	23	[10;33]	°C	8
T_{sc}	shape parameter of infection response to temperature	2.2e^{-3}	2.2e^{-3}		-	8

Horizontal lines separate the parameters related to canopy, pathogen and environment. Columns "stem" and "grid" indicate the value of parameters for the specific use-cases. Column "uncertainty" indicates the range of selected parameters in sensitivity analysis (in % when the meaning of the parameter is different in both models).
doi:10.1371/journal.pone.0049406.t002

doi:10.1371/journal.pone.0049406.t002

CANOPY DEVELOPMENT (UNIT CONSTRUCTION SUB-MODEL)

In plant models, time scale is frequently coupled with temperature as most physiological processes are temperature-dependent. The concept of thermal time [22] implies that time, as sensed by plants, elapses more rapidly at higher than at lower temperatures in a given range of temperatures [23] and is stopped at a base temperature. It is convenient as dates or rates expressed in thermal time are constant between environments. The main development stages for the crop (emergence, flowering, maturity) are expressed as a sum of degree-days (mean air temperature). As a consequence, calendar time (, day) and thermal time (°C) were defined as different mathematical notations.

In the Unit Construction model, the development process of the canopy was represented by the creation of functional units. This structural aspect is distinct from the functioning of each unit. Two pathosystems were taken as example to test the model genericity: (1) the pisum-aschochyta system [20], where the description of epidemics combines a local spore dispersal (splashing, stem-flow) and organ scale knowledge and (2) the potato late blight system that with long distance wind-dispersal between individualized plants [24]. Two corresponding methods, respectively named "stem" and "grid" are presented here to generate these contrasted canopy archetypes.

1. Stem method is used to simulate pathosystems presenting homogeneous canopies where epidemics follow a vertical dynamic. In this method, a single plant is represented as a succession of growing stem segments. The hypothesis of an homogeneous crop in the plot (mean plant) allowed aggregation at crop scale. The horizontal plot variability coming from soil, climate or primary inoculum repartition was not considered. In the model, functional units are created sequentially and recursively linked together to mimic stem growth. Thus, each functional unit represents a layer of organs having a similar physiological age over the whole canopy. The dynamic construction of the units was implemented with a finite state automaton (modelled with an Hortel statechart [25], a variant of the UML statechart [26]). In the statechart presented in figure 2, states correspond to units and the transitions between states are conditioned by thermal time accumulation. Different phyllochron values were defined to simulate preformed stem segments (is a vector). The unit construction process stops at flowering time ( ), determined at crop level. This method also defines the relationship between units: each unit is connected to its upper neighbour (linear graph).
2. Grid method is used to simulate patchy epidemics, with each unit representing an individual plant. In this case, all units are created simultaneously at the start of the simulation and are linked according to a network of connection. This global network is build from an a priori on the neighbourhood of individual units about both the pathogen dispersal process and the crop structure. This neighbourhood, relatively to individual units, is defined as a deterministic parameter (graph). Examples of possible individual neighbourhoods are presented in figure 3. The strength of the dispersal process was related to the number of plants considered in theses graphs (i.e graphs A vs B and D vsE). Constraints on range or direction of the dispersal process are represented by defining

asymmetric graphs (D, E, F). For example, graph E is used to simulate long-range dispersal events occurring at a low frequency. Graphs C and F are defined to represent row-structured crops, with a weak inter-row connection for graph F. At the crop scale, this theoretical network of unit connections is parameterized by the adjacency matrix of the corresponding graph. The outward edges of each node are weighted to adjust dispersal rate in relation to the leaf area development of the emitting unit (porosity attribute, cf. eq. 6).

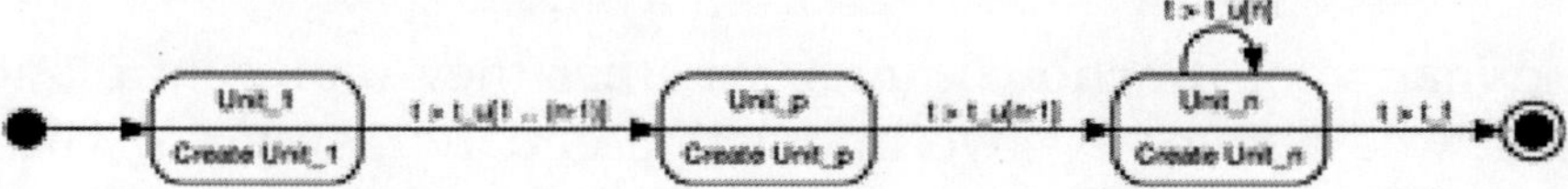

Figure 2 Hortel statechart for the "stem" method.

Hortel statechart for semi-determinate development of plant structure. The conditions for building a new unit (i.e thermal time above a threshold) are specified in transition between states and represents phyllochrons ( ) or crop flowering time ( ).

doi:10.1371/journal.pone.0049406.g002

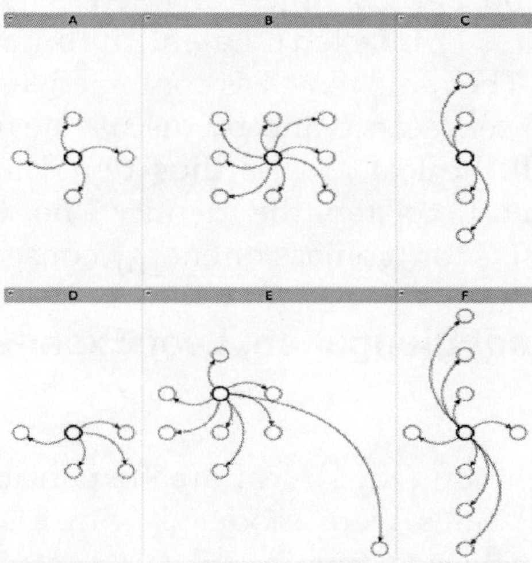

Figure 3 Individual neighbourhood for the grid method.

Nodes are individual plants and edges are pathogen dispersal processes. Local dispersion graphs are defined from a individual unit point of view (bold circle) and are used to infer the global network of connections between all functional units at the crop scale.

doi:10.1371/journal.pone.0049406.g003

INDIVIDUAL DYNAMICS FOR PLANT AND PATHOGEN GROWTH (FUNCTIONAL UNITSSUB-MODEL)

The dynamic of functional units, whether they represent plants or organs, was defined by two interacting equation systems. The first one is related to healthy plant functioning and the second depicts disease development on leaf tissues. These two systems include terms that depend on state variables from linked functional units, allowing to model disease development at the canopy scale.

Functional units were mainly described by leaf area because it is a driving variable for light interception (linked to crop productivity) and a physical support for pathogen development. The units also included height and porosity attributes. As in SEIR family models [4], [27], total leaf area for each functional unit is divided into distinct compartments. From the plant point of view, we defined expanded ( ), senescent ( ) and photosynthetically active ( ) leaf area. For the pathogen, the healthy ( ), infected (, Latent), infectious ( ) and removed ( ) compartments. The removed class combined leaf area destruction caused by natural senescence and by disease development. If there is no infection, all the leaf area is thus removed through natural senescence. At each time-step, the identity holds to maintain leaf area conservation (as for biomass or energy conservation).

Dynamics of Plant Elongation, Leaf Expansion, and Senescence

Stem height elongation (eq. 1), leaf area expansion (eq. 2) and leaf senescence (eq. 3) rates were modeled with a symmetric logistic equation driven by thermal time and 3 parameters. This formalism is frequently used at the crop scale in ecophysiology. It was previously adapted at organ-scale [28]. This approach is similar to ours when considering an analogy between organs and functional units.

1. Emax;Zmax the value of asymptote, representing the potential of the elongation ( ), expansion or senescence ( ) process whether a whole plant (total leaf area, plant height) or an organ (phytomer area, internode length) is considered. These parameters are measurable on plants growing in healthy conditions and without environmental stress.
2. te; ts; tz the value of the abscissa at the inflection point, representing thermal-time at half the value of or . Their value was estimated through optimization. We stated that , meaning that elongation is synchronized with expansion.
3. ke; ks; kz the value of the slope at the inflection point. This value is supposed to be constant among functional units (different between processes) and was optimized.

This model relies on two hypotheses: (1) the disease development does not affect the plant growth rate ( ) nor the potential size ( ), (2) there is no specific feedback of the disease on natural leaf senescence (is disease-independent).

Dynamic of Leaf Area in Disease Condition

The diffusion of pathogenic agents (i.e. spores) was not modelled explicitly and is approximated by rates of conversion between the different leaf area compartments. The aim of the pathogen model (eq. 4) was to merge a compartmental model of leaf injury and the host plant growth. This model was adapted from previously developed model on the powdery mildew - vineyard system[29]. In our case, two processes were refined: (1) healthy area was dependent of host growth and was simulated with a basic crop model and (2) both pathogen-induced and natural senescence ( ) contributed to the removal of leaf area from the system. It was also assumed that the natural senescence process removed leaf area independently from its health status (infected or infectious). It was a way to model disease resistance via avoidance, as leaf area can be removed before it became infectious. (4)

Disease Initiation

In the model, the primary infection event occurs at a pre-defined time (at day) and a fraction of healthy area ( ) is converted into the infected state. Depending on the epidemic archetype, either all units (for the "Stem" model) or a fixed proportion of the existing units at the time of the infection ("Grid" model) are affected. Disease's development. The dynamic for healthy area ( ) is given by the host growth ( ) minus the area infected from a local source minus the area infected from the neighbourhood minus the area that became senescent. In consequence, two parameters acting as rates of infection drive the disease progression in the unit (auto-deposition,) and between neighbouring units (allo-deposition,). The other terms in equation 4 are either regulation function ( ) or normalizing terms (area in the compartment divided by total area). The disease development is also a function of a latency period ( ) and an infectious period ( ). For computing simplicity, these periods are not simulated as a time-lag model, meaning that a fraction of infected area becomes infectious from day one after infection (with a rate of).

Disease regulation by crop properties

Receptivity, namely the level of ontogenic resistance of the plant tissues, affects local and distant infection rates. Equation 5 presents an example of decreasing ontogenic resistance with the unit aging and showing an increase in the rate of infection on the onset of leaf senescence ( ). Ontogenic resistance was modeled as a pathosystem-specific function [30] as tissues could get less resistant to infection with aging (pisum-aschochyta [31], potato late blight [32]) or, conversely, could become nearly immune in older organs (vitis-powdery mildew [33]). Biotrophic or necrotrophic behaviours were partly reproduced by this function.

The attribute of unit porosity integrates the architectural characteristics (e.g leaf area orientation, leaf surface properties) that are assumed to impact disease transmission and which are not explicitly modeled. Porosity is approximated (Eq. 6) by leaf area density (area/height) weighted by parameter . This regulation through porosity impacts the rate of infection from neighbouring units.

Canopy Architecture Impact on Disease Transmission

Predicting the population dynamics from the behaviour of individuals is a challenge when modeling epidemics [34]. To face this challenge, we formulated two hypotheses: (1) the individual-based compartment models (Eq. 4 and 7) are functions of crop-scale attributes, to make individual dynamics sensitive to global conditions and (2) the number of connections between units depends on the pathogen dispersal mode. The outgoing infection from a unit is modeled as multiplicative effect of infected area and unit porosity ( ). For each unit, the incoming infection ( ) is modeled as the sum of neighbouring emissions (cf. fig. 3.). Input and output disease flows are not synchronized: emissions are computed at the next time-step (the day after) in the target units. (7)

IMPACT OF ENVIRONMENTAL STRESS ON PLANT OR PATHOGEN DYNAMICS (ENVIRONMENT SUB-MODEL)

The Crop Climate model aimed to compute environmental impacts on pathogen development. At this step, only the temperature effect on disease development is modeled (Eq. 8). This factor was introduced as a multiplicative term that modulates disease transmission rates within and between units (Eq. 4). This bell-shape function depends on the crop mean temperature ( ) and two parameters: the optimum temperature for pathogen growth ( ) and the width of the response at this optimum temperature ( ). The parameter was defined as a normalizing term to allow dimensional consistency.

Software Implementation and Data Analysis

Software Implementation

The model development and simulations rely on the VLE software platform (Virtual Laboratory Environment) [35] and the INRA RECORD platform [36]. This platform implements a mathematical specification (DEVS, Discrete EVent system Specification [37]) that was originally developed for research in modeling and simulation independently of the domain of application. The main feature of the VLE platform is that it proposes software sub-formalisms between the abstract DEVS specifications and common mathematical formalisms in environmental sciences (e.g ODE, Hortel statechart, cellular automatons). Consequently, most of the formal model was implemented by using these extensions rather than manipulating underlying DEVS formalism and making it possible to integrate heterogeneous formalisms in the same model. Dynamic creation of model structure. Following the class/object paradigm, the Functional units model is defined as a class model whose instantiation depends on the model graph structure. Depending on the pathosystem, instantiation is static (at plant emergence) or dynamic during simulation (following the plant development); these new units interact according to the specification of the connection network. This allowed dissociating the biological representation from its consequence on the graph and simulation.

Data Analysis

The sensitivity analysis and data visualization was performed with R core software [38] with R packages sensitivity and ggplot2 [39]. Global sensitivity analysis aims at explaining the variability of model outputs with respect to the input parameter uncertainties. It is based on intensive use of computer simulations. After setting an experimental design of parameter values, model outputs are evaluated for each parameter set and sensitivity indices are computed. Various methods of global sensitivity analysis exist [40]. To avoid the model evaluations of a full factorial design (10 parameters with each m modalities), we used the extended-FAST method[41] (size , with) to perform the sensitivity analysis. It allows to evaluate a first-order sensitivity index (influence of only one parameter on the output) and a total sensitivity index (influence of one parameter alone and in interaction with the others parameters). These two indices correspond, in the classical variance analysis, to the principal effect and the total (principal+interactions) effect.

DISCUSSION

Qualitative Model Evaluation

A strong hypothesis is that the model genericity is fully supported by the parameter set (cf. table 2) and the individual graph generating the global connection network (cf. figures 2 and 3). This hypothesis was tested here by running the model with two sets of parameters. The first set of parameter values characterizes a vertical epidemic in a homogeneously layered canopy whereas the second represents a spatial epidemic in a field (lattice of individual plants). These parameter sets will be later referred to as vertical or spatial epidemics. In our approach, the model has been built from a conceptual point of view to get gradually closer to the biological reality. In this sense, the parameter estimation on real data was not our priority in the development of the model. Nevertheless, some parameters are quite easy to estimate, especially those related to phenology as they are well studied for a number of crops (e.g. in pea[42], [43], potato [44], [45]). In the present work, parameters were not estimated with a numerical optimization method, but were defined so that the epidemic dynamics are meaningful according to the pathologists' knowledge.

The first step concerning the model evaluation was to reproduce real behaviours, those that pathologists observe in field conditions. The output variable used to evaluate the model and perform a sensitivity analysis is the proportion of leaf area removed by the sole effect of disease ( ), integrated over time (sowing to harvest) and space (units).

Ascochyta Blight on Pea

The stake was to reproduce "infection profiles", i.e. a varying disease score upon the canopy height, caused by the development of ascochyta blight on pea fields [20]. This observed behaviour is assumed to be a consequence of two processes: a frequent disease initiation on lower leaves (older and more receptive) and a local dispersion to upper nodes [31]. The bell shape of the disease profiles is interpreted as a gradient of age-based (ontogenic) resistance, as upper and younger leaves are more resistant to infection and lower leaves could become senescent before being fully infected. The model outputs are presented in figure 4. They display a classical behaviour as the disease starts developing firstly on lower nodes, moving upwards with time (figure 4, lower graph) [46]. Even if individual unit dynamics are rarely observed in field, the upper section of figure 4 indicates that after the beginning of infection ( ), the system was perturbed to time about before returning to a healthy state. This result is interpreted by the combination of two effects: a high host growth rate (caused by a favorable oceanic climate) and a weak disease transmission rate between plant nodes ( ). Refining the model prediction at this scale will require additional experiments on controlled conditions (e.g. on leaf receptivity, [31]).

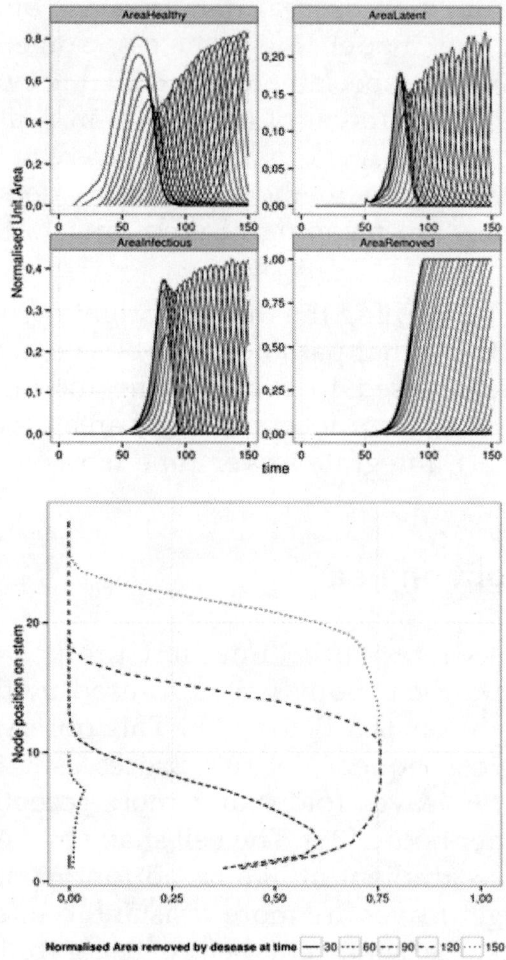

Figure 4 Vertical epidemics: disease development in a homogeneously layered canopy.

The four upper graphs show individual dynamics of functional units, with each graph corresponding to a state of infection. The output variables are relative to the total leaf area in the unit. The lower graph presents the evolution of "infection profiles", namely the severity of the disease in relation to the height (stem nodes) in the crop.

doi:10.1371/journal.pone.0049406.g004

Potato Late Blight

In this case, the model parameters intend to represent a spatial epidemic, where random primo-infection generates primary foci which in turn infect plants in their neighbourhood [24]. It is not clear how to infer the connection network from the real system. On one hand, observation of patches of infection means a short-range diffusion process (new infections are more likely to occur near a previously infected plant). On the other hand, a long-range dispersal is often observed [47]. Thus, the simulations showed in figure 5 aim at exploring the sensitivity of the epidemic process to changes in size or range in the dispersal neighbourhood. Four examples were chosen from graphs in figure 3. The asymmetry of some graphs, used as connection network, provide interesting heterogeneity at the canopy scale. This parameterization could represent epidemics in heterogeneous fields (e.g. prevailing winds, asynchronous plant emergence). These results illustrate the plasticity we achieved in simulation rather than a formal sensitivity analysis on this connection network.

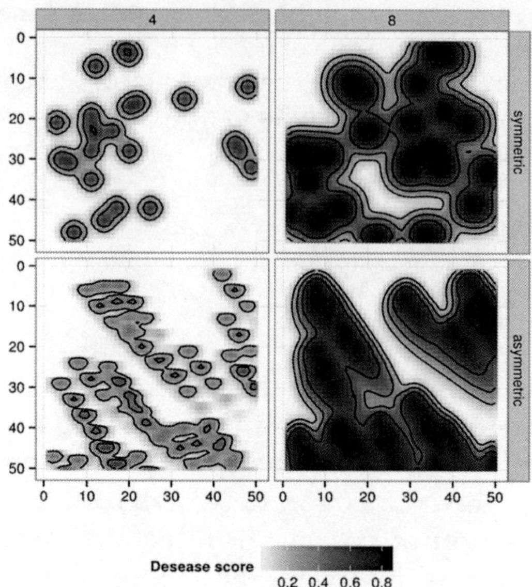

Figure 5. Spatial epidemics: disease development in a lattice of individualised plants.

The four maps show the disease severity (grey scale) on grids of 50×50 plants at the end of simulation. Disease initiation was done at random on 1 percent of total cell number for both cases. These maps differed in the parameterization of the connection network between plants (cf. figure 3): upper maps resulted from symmetrical graphs (i.e. graphs A and B in figure 3), lower maps resulted from asymmetrical graphs (D and E); right and left maps resulted from a variation in the number of plants considered for local dispersion (4 and 8).

doi:10.1371/journal.pone.0049406.g005

With our model, a quantitative proportion of injured leaf area is used as output data, as pest injuries are usually observed on quantitative scales. Observations to discriminate between pathogen-induced or senescence-induced removal of leaf area is a very difficult task, if observation is not performed continuously. This modeling approach permits to estimate the respective proportion of leaf area removal caused by these two processes. Protocols for characterization of pest injuries in the field [48] should be addressed in order to evaluate the predictive quality of our modeling.

SENSITIVITY ANALYSIS

We have seen in the former section that it was possible to reproduce observed real behaviors using specific sets of parameter values. Sensitivity analysis can be useful to determine which parameters are the most influential on the variability of the model output, which are those having no influence (precise knowledge on these parameters may not be not necessary), which are those interacting together. It can help the modeler to check if the model respects the a prioriknowledge on relative effect of parameters and their interactions. This analysis can also help set research priorities. For example, to help decide which experiments are needed to improve the precision on the estimation of some parameters or to identify regions of the parameter space with a high (or low) model output which can be useful on a prospective point of view.

A global sensitivity analysis of the model was conducted to further apprehend its behavior (cf.figure 6). The impact of 10 parameters on disease severity at the crop scale was assessed for the two pathosystems (ascochyta blight on pea and potato late blight) on two contrasted climatic environments (Oceanic - Rennes, Western France and Mediterranean - Montpellier, South of France). The range of variation for the input parameters was adjusted to keep a biological meaning. We assumed a uniform distribution within this range (figure 6, column "uncertainty"). The Extended-FAST method [41] was used to explore parameter space and compute sensitivity indices.

Whatever the model, the three most influential parameters on disease severity are the disease initiation day, the latency period and the onset of senescence. All three parameters are time-related. This analysis points out that avoidance strategy strongly explains disease severity: whether this avoidance came from a low exposure (late inoculum arrival), a slow disease dispersal (long latency period) or a de-synchronisation of plant and pathogen cycle (early plant senescence). The high impact of the disease initiation day parameter on the severity suggests that the control of primary inoculum production is a key point for crop protection. This point should be addressed in connection with management strategies, i.e. interaction between tillage and crop sequence to manage infected stubble, [49].

As expected, estimated sensitivity indices highlight the strong impacts of both the network transmission and senescence parameters. The impact of senescence on severity is higher in spatial than in vertical epidemics. This is interpreted as the consequence of the synchronisation of host development in the spatial model. Interestingly, all parameters interacted more tightly with each other in spatial epidemics (larger total index).

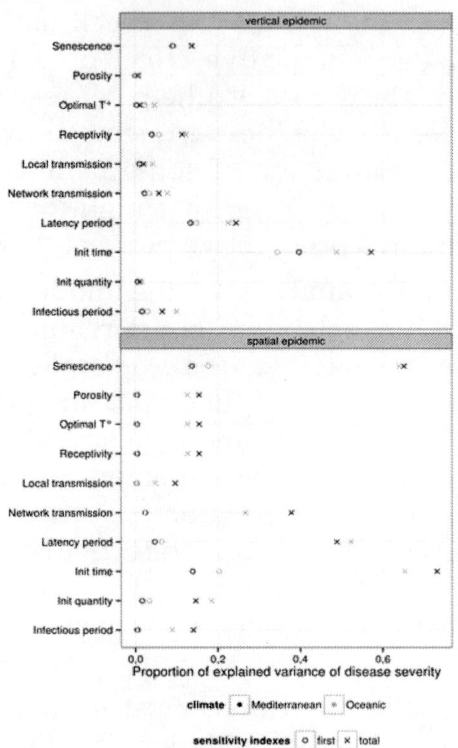

Figure 6 Sensitivity analysis.

Impact of 10 parameters from table 2 on disease severity for the two patho-systems: top, ascochyta blight on pea, and bottom, potato late blight. Main effect (circle) and total effect (cross) for each parameters are estimated with Extended FAST method [41] on two contrasted climatic conditions: Mediterranean (black) and oceanic (grey).

doi:10.1371/journal.pone.0049406.g006

Both models are sensitive to climate (temperature) and show contrasted response. In Mediterranean climate, the disease severity is globally more sensitive variation of model parameters, for the two types of epidemics. In the warmer climate, disease initiations concomitant with a senescent crop (resulting from a hastened crop development) appeared to be more frequent and thus a higher epidemic variance was observed. For the spatial epidemic case, the impact of host ontologic resistance (receptivity) and optimal temperature for pathogen development shifted from none to significant with the change in environmental conditions. This indicates that the model is able to reproduce interesting host pathogen environment interactions and could be useful to better understand some mechanisms of plant epidemics.

CONCLUSIONS

As stated before, the developed model lies in the research domain. It needs more modeling iterations with phytopathologists to be able to produce outputs that can be used for agricultural applications. Nevertheless, the modeling process raises interesting questions that could be scrutinised with experimental approaches: the disease impact on plant senescence, previously viewed as an additional process to the model, was finally represented as a simple competition for leaf area. Simulation plays a great role to help grasping the functioning of agricultural systems where direct observations are difficult and tedious, providing a sound balance between real and virtual experiments.

About Conceptual Development

A consequent part of the model adaptation to specific systems was achieved through the structure of unit network. As its impact cannot be directly evaluated through classical sensitivity analysis, additional virtual experiments would be necessary to develop an intuitive understanding of network-based epidemics and the effect of network structure [13]. Thereby, to simulate systems with a high number of units and a nearly-complete connection graph (e.g. vineyards) will be highly demanding in computing time: statistics on connection network on real pathosystems would help reduce this uncertainty. A first step, to improve the present model, is to focus on the climatic environment at the local scale (microclimate or phylloclimate [50]) as this local heterogeneity could probably explain finer host pathogen environment interactions. The modular model conception makes this improvement easy to take into account. Additionally, environmental stresses are most probably not sensed independently by plant (or organs). An interesting model combining the action of temperature and wetness on pathogen response, proposed by Duthie [51], could be considered to integrate the joint effect of stresses for a range of foliar pathogens. Cropping practices (e.g. pruning, lodging) that dramatically affect the crop structure are not yet included in the present model and should improve its relevance for practical applications.

About Practical Development

The modeling of two types of epidemics within one single model structure was a first step demonstrating that the defined objects (functional units and a network of connection) had sufficient flexibility to mimic different crop/pathogen systems. Because we focused on comparative epidemiology rather than on quantitative evaluation, operational conclusions are still out of the scope of the model, and are therefore not presented in this paper. Very few studies have integrated canopy architecture and functioning as a way to control disease [52],[53]. Among these studies, the crop-pathogen system is mainly modeled with functional-structural plant framework (FSPM) including 3D representation of the canopy. Using a coarser-grained and less computationally intensive model makes uncertainties (climate variability and change, diverse production systems) easier to analyse within a simulation scheme. Biological complexity is thus easier to integrate and assess [54].

Integrative biology can be seen not only as integration between scales but also between the different points of view (agronomists, pathologists and mathematicians). This work also highlighted the role of software simulation platforms as tools to integrate the knowledge coming from these different fields [55].

ACKNOWLEDGMENTS

The authors warmly thank agronomists and phytopathologists implicated in the project for their fruitful collaboration. The authors also specially thanks Jean-Noël Aubertot for his helpful review of the manuscript.

AUTHOR CONTRIBUTIONS

Conceived and designed the experiments: PC RF ML. Performed the experiments: PC GQ. Analyzed the data: PC RF. Contributed reagents/materials/analysis tools: PC GQ RF. Wrote the paper: PC RF GQ. Designed the modeling and simulation platform: GQ.

REFERENCES

1. Oerke E (2006) Crop losses to pests. The Journal of Agricultural Science 144: 31–43. doi: 10.1017/s0021859605005708
2. Boote K, Jones J, Mishoe J, Berger R (1983) Coupling pests to crop growth simulators to predict yield reductions. Phytopathology 73: 1581–1587. doi: 10.1017/s0021859605005708
3. Papaïx J, Goyeau H, Du Cheyron P, Monod H, Lannou C (2011) Inuence of cultivated landscape composition on variety resistance: an assessment based on wheat leaf rust epidemics. New Phytologist 191: 1095–1107. doi: 10.1111/j.1469-8137.2011.03764.x
4. Madden L (2006) Botanical epidemiology: some key advances and its continuing role in disease management. Plant disease epidemiology: facing challenges of the 21st Century 3–23. doi: 10.1017/s0021859605005708
5. Aubertot J, West J, Bousset-Vaslin L, Salam M, Barbetti M, et al.. (2006) Sustainable strategies for managing Brassica napus (oilseed rape) resistance to Leptosphaeria maculans (phoma stem canker), Springer, chapter Improved resistance management for durable disease control: a case study of phoma stem canker of oilseed rape (Brassica napus).
6. Quilot B, Kervella J, Genard M, Lescourret F (2005) Analysing the genetic control of peach fruit quality through an ecophysiological model combined with a QTL approach. Journal of Experimental Botany 56: 3083–3092. doi: 10.1093/jxb/eri305

7. Chenu K, Chapman S, Tardieu F, McLean G, Welcker C, et al. (2009) Simulating the yield impacts of organ-level quantitative trait loci associated with drought response in maize: A "gene-tophenotype" modeling approach. Genetics 183: 1507. doi: 10.1534/genetics.109.105429
8. Hau B (1990) Analytic models of plant disease in a changing environment. Annual Review of Phytopathology 28: 221–245. doi: 10.1146/annurev.py.28.090190.001253
9. Vanderplank J (1982) Host-pathogen interactions in plant disease. Academic press, New York.
10. Jeger M (1982) The Relation Between Total, Infectious, and Postinfectious Diseased Plant Tissue. Phytopathology 72: 1185–1189. doi: 10.1094/phyto-77-1185
11. Jeger M (1986) The potential of analytic compared with simulation approaches to modeling in plant disease epidemiology. Plant Disease Epidemiology: Population Dynamics and Management 1: 255–281. doi: 10.1094/phyto-77-1185
12. Willocquet L, Savary S (2004) An epidemiological simulation model with three scales of spatial hierarchy. Phytopathology 94: 883–891. doi: 10.1094/PHYTO.2004.94.8.883
13. Keeling M, Eames K (2005) Networks and epidemic models. Journal of the Royal Society Interface 2: 295. doi: 10.1094/phyto-77-1185
14. Gilligan CA, van den Bosch F (2008) Epidemiological models for invasion and persistence of pathogens. Annual Review of Phytopathology 46: 385–418. doi: 10.1094/phyto-77-1185
15. Tardieu F (2010) Why work and discuss the basic principles of plant modelling 50 years after the first plant models? Journal of Experimental Botany 61: 2039. doi: 10.1093/jxb/erq135
16. Vos J, Evers J, Buck-Sorlin G, Andrieu B, Chelle M, et al. (2010) Functional-structural plant modelling: a new versatile tool in crop science. Journal of Experimental Botany 61: 2101. doi: 10.1093/jxb/erp345
17. Ferrandino F (2008) Effect of Crop Growth and Canopy Filtration on the Dynamics of Plant Disease Epidemics Spread by Aerially Dispersed Spores. Phytopathology 98: 492–503. doi: 10.1094/PHYTO-98-5-0492
18. Robert C, Fournier C, Andrieu B, Ney B (2008) Coupling a 3d virtual wheat (triticum aestivum) plant model with a septoria tritici epidemic model (septo3d): a new approach to investigate plant–pathogen interactions linked to canopy architecture. Functional Plant Biology 35: 997–1013. doi: 10.1071/fp08066

19. Wilson P, Chakraborty S (1998) The virtual plant: a new tool for the study and management of plant diseases. Crop Protection 17: 231–239. doi: 10.1071/fp08066
20. Le May C, Ney B, Lemarchand E, Schoeny A, Tivoli B (2008) Effect of pea plant architecture on spatiotemporal epidemic development of ascochyta blight (Mycosphaerella pinodes) in the field. Plant Pathology doi: 10.1111/j.1365-3059.2008.01947.x
21. Urban D, Keitt T (2001) Landscape connectivity: a graph-theoretic perspective. Ecology 82: 1205–1218. doi: 10.1111/j.1365-3059.2008.01947.x
22. Trudgill D, Honek A, Li D, Van Straalen N (2005) Thermal time–concepts and utility. Annals of Applied Biology 146: 1–14. doi: 10.1111/j.1365-3059.2008.01947.x
23. Granier C, Tardieu F (1998) Is thermal time adequate for expressing the effects of temperature on sunower leaf development? Plant, Cell & Environment 21: 695–703. doi: 10.1111/j.1365-3059.2008.01947.x
24. Scherm H (1996) On the velocity of epidemic waves in model plant disease epidemics. Ecological Modelling 87: 217–222. doi: 10.1111/j.1365-3059.2008.01947.x
25. Harel D (1987) Statecharts: A visual formalism for complex systems. Science of Computer Programming 8: 231–274. doi: 10.1016/0167-6423(87)90035-9
26. OMG Unified Modeling LanguageTM (OMG UML), Superstructure Version2.2 (2009) Available at http://www.omg.org/spec/UML/2.2/Superstructure. Version 2.2 is a minor revision to the UML 2.1.2 specification. It supersedes formal/2007-11-02.
27. Scherm H, Ngugi H, Ojiambo P (2006) Trends in theoretical plant epidemiology. European journal of plant pathology 115: 61–73. doi: 10.1016/0167-6423(87)90035-9
28. Lizaso JI, Batchelor WD, Westgate ME (2003) A leaf area model to simulate cultivar-specific expansion and senescence of maize leaves. Field Crops Research 80: 1–17. doi: 10.1016/0167-6423(87)90035-9
29. Burie J, Langlais M, Calonnec A (2011) Switching from a mechanistic model to a continuous model to study at different scales the effect of vine growth on the dynamic of a powdery mildew epidemic. Annals of Botany 107: 885. doi: 10.1093/aob/mcq233
30. Develey-Rivière M, Galiana E (2007) Resistance to pathogens and host developmental stage: a multifaceted relationship within the plant kingdom. New Phytologist 175: 405–416. doi: 10.1111/j.1469-8137.2007.02130.x

31. Richard B, Jumel S, Rouault F, Tivoli B (2011) Inuence of plant stage and organ age on the receptivity of pisum sativum to mycosphaerella pinodes. European Journal of Plant Pathology In press doi: 10.1007/s10658-011-9882-3
32. Carnegie S, Colhoun J (1982) Susceptibility of potato leaves to phytophthora infestans in relation to plant age and leaf position. Journal of Phytopathology 104: 157–167. doi: 10.1007/s10658-011-9882-3
33. Doster M, Schnathorst W (1985) Effects of leaf maturity and cultivar resistance on development of the powdery mildew fungus on grapevines. Phytopathology 75: 318–321. doi: 10.1007/s10658-011-9882-3
34. Kleczkowski A, Gilligan C, Bailey D (1997) Scaling and spatial dynamics in plant-pathogen systems: from individuals to populations. Proceedings of the Royal Society B: Biological Sciences 264: 979. doi: 10.1007/s10658-011-9882-3
35. Quesnel G, Duboz R, Ramat E (2009) The Virtual Laboratory Environment – An operational framework for multi-modelling, simulation and analysis of complex dynamical systems. Simulation Modelling Practice and Theory 17: 641–653. doi: 10.1016/j.simpat.2008.11.003
36. Bergez J, Chabrier P, Gary C, Jeuffroy M, Makowski D, et al. (2012) An open platform to build, evaluate and simulate integrated models of farming and agro-ecosystems. Environmental Modelling & Software doi: 10.1016/j.simpat.2008.11.003
37. Zeigler B, Praehofer H, Kim T (2000) Theory of modeling and simulation : Integrating discrete event and continuous complex dynamic systems, volume 100. Academic press New York.
38. R Development Core Team and others (2010) R: A language and environment for statistical computing. R Foundation for Statistical Computing Vienna Austria.
39. Wickham H (2009) ggplot2. Wiley Interdisciplinary Reviews: Computational Statistics.
40. Saltelli A, Chan K (2000) Sensitivity Analysis. Wiley.
41. Saltelli A, Tarantola S, Chan K (1999) A Quantitative Model-Independent Method for Global Sensitivity Analysis of Model Output. Technometrics 41: 39–56. doi: 10.2307/1270993
42. Turc O, Lecoeur J (1997) Leaf Primordium Initiation and Expanded Leaf Production are Coordinated through Similar Response to Air

Temperature in Pea (Pisum sativumL.). Annals of Botany 80: 265. doi: 10.2307/1270993

43. Roche R, Jeuffroy M, Ney B (1998) A model to simulate the final number of reproductive nodes in pea (pisum sativuml.). Annals of Botany 81: 545. doi: 10.2307/1270993
44. Jefferies R (1989) Water-stress and leaf growth in field-grown crops of potato (solanum tuberosum l.). Journal of experimental botany 40: 1375. doi: 10.2307/1270993
45. Fleisher D, Timlin R, Kim D, Reddy S, Vangimalla R (2006) Approaches to modeling potato leaf appearance rate. Agronomy Journal 98: 522. doi: 10.2134/agronj2005.0136
46. Le May C, Schoeny A, Tivoli B, Ney B (2005) Improvement and validation of a pea crop growth model to simulate the growth of cultivars infected with Ascochyta blight (Mycosphaerella pinodes). European Journal of Plant Pathology 112: 1–12. doi: 10.2134/agronj2005.0136
47. Johnson D, Alldredge J, Hamm P, Frazier B (2003) Aerial photography used for spatial pattern analysis of late blight infection in irrigated potato circles. Phytopathology 93: 805–812. doi: 10.1094/PHYTO.2003.93.7.805
48. Deytieux V, Cellier V, Agerberg J, Armand J, Boll R, et al.. (2010) Quantipest, a collaborative platform to help quantify pest populations and injuries in the field. In: ENDURE International Conference, 23–24 November. Paris, France.
49. Schneider O, Roger-Estrade J, Aubertot J, Doré T (2006) Effect of seeders and tillage equipment on vertical distribution of oilseed rape stubble. Soil and Tillage Research 85: 115–122. doi: 10.2134/agronj2005.0136
50. Chelle M (2005) Phylloclimate or the climate perceived by individual plant organs: What is it? How to model it? What for? New Phytologist 166: 781–790. doi: 10.1111/j.1469-8137.2005.01350.x
51. Duthie J (1997) Models of the response of foliar parasites to the combined effects of temperature and duration of wetness. Phytopathology 87: 1088–1095. doi: 10.1094/PHYTO.1997.87.11.1088
52. Pangga I, Hanan J, Chakraborty S (2011) Pathogen dynamics in a crop canopy and their evolution under changing climate. Plant Pathology 60: 70–81. doi: 10.1111/j.1365-3059.2010.02408.x
53. Ando K, Grumet R, Terpstra K, Kelly J (2007) Manipulation of plant architecture to enhance crop disease control. CAB Review doi: 10.1111/j.1365-3059.2010.02408.x

54. Hammer G, Cooper M, Tardieu F, Welch S, Walsh B, et al. (2006) Models for navigating biological complexity in breeding improved crop plants. Trends in Plant Science 11: 587–593. doi: 10.1016/j.tplants.2006.10.006
55. Affholder F, Tittonell P, Corbeels M, Roux S, Motisi N, et al. (2012) Ad hoc modeling in agronomy: What have we learned in the last 15 years? Agronomy Journal 104: 735. doi: 10.2134/agronj2011.0376

Chapter 8

EPISTATIC ASSOCIATION MAPPING IN HOMOZYGOUS CROP CULTIVARS

[1]Hai-Yan Lü, [1]Xiao-Fen Liu, [1]Shi-Ping Wei, [1]Yuan-Ming Zhang

[1]Section on Statistical Genomics, State Key Laboratory of Crop Genetics and Germplasm Enhancement, Nanjing Agricultural University, Nanjing, Jiangsu, China

[2]College of Information and Management Science, Henan Agricultural University, Zhengzhou, Henan, China

ABSTRACT

The genetic dissection of complex traits plays a crucial role in crop breeding. However, genetic analysis and crop breeding have heretofore been performed separately. In this study, we designed a new approach that integrates epistatic association analysis in crop cultivars with breeding by design. First, we proposed an epistatic association mapping (EAM) approach in homozygous crop cultivars. The phenotypic values of complex traits, along with molecular marker information, were used to perform EAM. In our EAM, all the main-effect quantitative trait loci (QTLs), environmental effects, QTL-by-environment interactions and QTL-by-QTL interactions were included in a full model and estimated by empirical Bayes approach. A series of Monte Carlo simulations was performed to confirm the reliability of the new method. Next, the information from all detected QTLs was used to mine novel alleles for each locus and to design elite cross combination. Finally, the new approach was adopted to dissect the genetic basis of seed length in 215 soybean cultivars obtained, by stratified random sampling, from 6 geographic ecotypes in China. As a result, 19 main-effect QTLs and 3 epistatic QTLs were identified, more than 10 novel alleles were mined and 3 elite parental combinations, such as Daqingdou and Zhengzhou790034, were predicted.

INTRODUCTION

Germplasm resources play crucial roles in genetics, evolution and breeding, by forming the physical foundation of the study of genetic diversity [1]–[3], fueling much evolutionary research[4]–[6] and providing the raw material for breeders to produce new cultivars or to further improve the existing ones, due to the existence of many valuable genes in genetic resources[7]–[9]. The identification of valuable genes and markers associated with traits of interest will greatly increase the efficiency of plant breeding programs. However, these beneficial genes are largely unexplored due to the lack of appropriate statistical techniques. Meanwhile, as the complexity of the trait increase, breeding problems increase, for example, favorable alleles in exotic genetic resources are in unadapated genetic backgrounds and linked to other unfavorable alleles. This means that methods to utilize these favorable alleles in crop breeding also need to be further addressed. Accordingly, there is a critical need for in-depth study of methodologies for mining elite alleles in germplasm resources and for the utilization of these elite alleles in crop breeding.

During the past several decades, many attempts have been made to mine elite alleles for objective traits of interest. In early studies, many genes for qualitative traits in crop breeding were studied with morphological and biochemical approaches [10]–[13], and those for complex diseases in human genetics were identified by both sibling pair analysis [14]–[18] and pedigree analysis [19]–[21]. The introduction of molecular markers has facilitated the genetic association analysis of complex diseases in humans, animals and plants. Single-marker association analysis [22] and, later, genome-wide association study (GWAS) have been widely used in human genetics [23]. There has been substantial research of two aspects of GWAS: population structure [24]–[29] and mixed genetic models [30]–[32]. However, only one QTL was analyzed at a time in the above models. Likewise, although epistasis association analysis has been utilized in human genetics [33]–[37], all of the main genetic effects and gene interaction effects have not been simultaneously included in one genetic model. A full genetic model, including all the main and epistatic effects, could improve the power of QTL detection[38]–[41]. Several parameter estimation approaches such as LASSO [41], [42], empirical Bayes [43], and penalized maximum likelihood [38], [40] make this full genetic model possible. Therefore, epistasis association analysis with a full genetic model is feasible in crop germplasm resources.

In the past, most crop breeding methods were based on selection for observable phenotypes and breeding efficiency without markers is simply a function of heritability and choice of parental material. To date molecular markers have improved efficiency of selection largely for traits under simple genetic control and in specific conditions where marker selection is easier/cheaper than phenotypic selection [44]–[50]. However, this approach is only feasible for the improvement of one or several independent genes. If there are interactions among the objective genes, breeding strategy must be addressed by the incorporation of the epistasis[51], [52]. Carlborg and Haley [53] showed that epistasis is a common response to selection in breeding programs. Therefore, genetic interaction should be considered in crop breeding strategies.

One purpose of the genetic analysis of quantitative traits is to design a suitable breeding strategy, called breeding by design [54]. However, genetic analysis and crop breeding have traditionally been performed separately; for example, most genetic analyses exclusively use biparental crosses, but these are rarely used alone in commercial breeding. Therefore, the results of these biparental cross experiments have limited roles in breeding practice [55]–[57]. However, direct mapping of QTLs in natural populations, such as crop cultivars, is both economical and practical because the population being mapped is readily available, and the identified QTLs are directly applicable [31]. The purpose of this study was to develop an epistatic association mapping (EAM) approach in homozygous crop cultivars. We described detailed genetic and statistical models of epistasis association analysis in crop cultivars. All the parameters were estimated using the empirical Bayes approach. Our methods were confirmed by real data analysis in soybean and by a series of Monte Carlo simulation experiments.

RESULTS

Phenotypic variation

We measured seed length in 215 soybean cultivars. The minimum, maximum, average, median, standard deviation, coefficient of variation, skewness and kurtosis values were 5.30, 11.85, 7.94, 7.86, 0.99, 12.43, 0.61 and 0.91, respectively. Results from ANOVA showed that there is significant difference among cultivars ($P<10^{-4}$) and there are no significant differences between years ($P = 0.192$) and among cultivar × year interactions ($P = 0.328$). This means that in the cultivar population, there is a large amount of genetic variation, which exhibits a continuous normal distribution (**Fig. 1**).

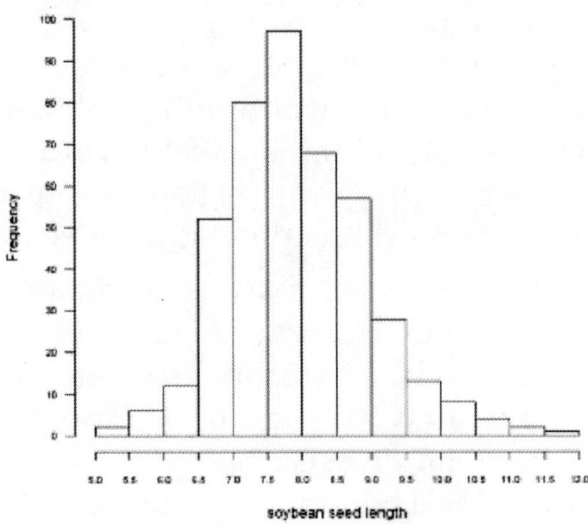

Figure 1. Frequency distribution for soybean seed length.

doi:10.1371/journal.pone.0017773.g001

Epistasis association mapping

Two years of phenotypic observations, along with information on 134 SSR molecular markers, were used to dissect the genetic basis of seed length in soybean. In the full model, 9,180 effects needed to be estimated, 40 times larger than the sample size. We adopted a two-stage method [58]. Nineteen main-effect QTLs and 3 epistatic QTLs for seed length in soybean were detected by EAM (Table 1). All of these QTLs were nearly evenly distributed along the soybean genome, except for chromosomes H, J and L. Among these QTLs, the proportion of the total phenotypic variance was from 0.25% to 10.44% for main-effect QTLs and from 5.08% to 7.38% for epistatic QTLs, and each of 12 QTLs contributed greater than 5.0% of the variance. In addition, five loci were involved in epistatic interactions, and only one of these five (sat_342) had a significant main effect. This lack of main effects may create difficulties in detecting epistasis with other methods.

Table 1. Detected QTL for seed length in soybean cultivar population.

QTL	New method						Genome-wide association study		
	Chr.	Marker associated	Position (cM)	Variance*	LOD	r^2 (%)	F	P-value	$-\log_{10}P$
Main-effect	A1	satt382	36.43	0.5115	4.85	6.24	4.31	3.96E-7	6.40**
	A2	satt429	116.98	0.8199	2.61	1.06	7.12	1.53E-1	0.81
	B1	satt509	52.51	0.8428	1.99	2.30	4.67	3.89E-4	3.41
	B2	sat_342	20.31	0.8346	4.81	5.15	2.28	8.33E-3	2.08
	B2	satt534	87.90	0.7554	2.05	10.44	3.05	3.56E-5	4.50
	C2	sat_252	127.00	0.8962	4.85	5.19	3.73	1.90E-6	5.70
	D1b	sat_254	46.52	0.8709	4.12	3.65	4.24	1.27E-7	6.90**
	D1b	satt274	116.35	0.8003	6.93	8.45	10.87	2.27E-5	4.64
	D2	satt514	65.69	0.9659	6.33	5.32	2.05	1.31E-5	4.88
	D2	sat_355	97.39	0.9232	15.23	5.60	3.08	1.70E-6	3.74
	E	satt263	45.40	0.8592	3.67	3.30	3.71	1.17E-2	1.93
	F	satt506	135.12	0.7007	4.71	5.44	2.47	2.29E-3	2.64
	G	satt352	10.53	0.5307	3.37	7.06	1.74	3.46E-3	1.46
	G	AW183283	87.94	0.8122	3.77	7.30	6.38	1.86E-3	2.73
	I	sat_418	98.11	0.8047	6.24	8.35	7.64	2.98E-6	5.22
	K	satt441	46.30	0.8025	6.55	5.00	3.22	1.64E-7	6.90**
	M	sat_258	74.93	0.8695	2.58	4.82	3.57	5.61E-3	2.30
	N	satt022	102.06	0.5115	11.89	6.61	2.16	2.92E-3	2.53
	O	sat_274	107.58	0.9446	2.54	2.41	2.61	5.21E-4	3.28
Epistasis	B2 & C1	sat_342 & AW277661	20.31 & 74.79	0.5367	7.71	7.38	4.04	6.72E-6	5.17
	D1a & E	sat_158 & satt411	104.20 & 13.92	0.8944	3.06	5.08	3.74	3.67E-4	3.51
	D1b & E	satt459 & satt411	116.62 & 13.92	0.5224	3.61	6.61	6.73	1.33E-3	2.88

*: Calculated by [illegible] for main-effect QTL and [illegible] for epistatic QTL, where f is allelic frequency, a is allelic effect and n and m is the number of alleles at the ith and jth loci. The same is true for the later tables.
**: QTL identified by genome-wide association study with the critical value at the 0.05 level of significance determined by 1000 permutation experiments.
doi:10.1371/journal.pone.0017773.t001

doi:10.1371/journal.pone.0017773.t001

To compare the proposed approach with regular genome-wide association study (GWAS), the GWAS was used to analyze the above dataset. Results showed that three main-effect QTL, linked with markers satt382, sat_254 and satt441, respectively, were detected (**Fig. 2a**) and no significant environmental and epistatic interactions were identified (**Fig. 2**). These results are similar to those by the proposed approach in two aspects. First, the three main-effect QTLs detected by the GWAS are also identified by the proposed method. Second, no significant environmental interaction is detected by the above two approaches. However, there are some differences as well. The main difference is that the new approach can detect more main-effect and epistatic QTLs than the GWAS.

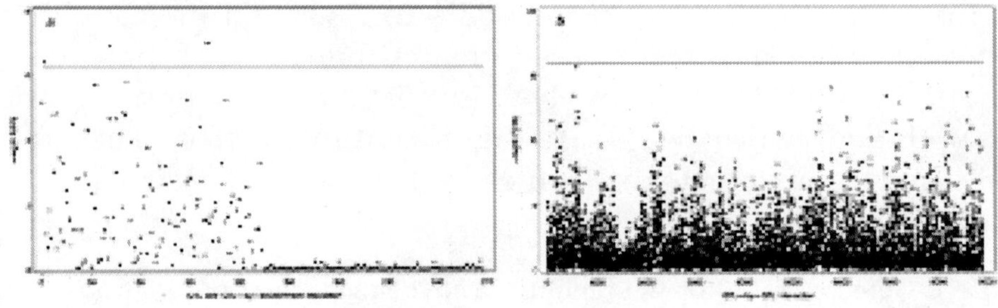

Figure 2. The $-\log_{10}P$ score profile of the soybean genome scan in the genome-wide association study for seed length in soybean.

(a) Main-effect QTL and QTL-by-environmental interaction, and (b) QTL-by-QTL interaction. The critical values at the 0.05 level of significance, indicated by horizontal line, were determined by 1000 permutation experiments.

doi:10.1371/journal.pone.0017773.g002

Mining elite alleles

The allelic effects of the cultivars were evaluated for all the identified loci for soybean seed length. The reduced model that includes the total mean, the population structure, all the identified loci and the residual error was a mixed model equation. In the reduced model, the allelic effects at each locus were estimated by a maximum likelihood approach. If we want to increase the trait value, we should take the allele with the largest positive effect per main-effect QTL as novel allele. If decreasing the trait value is our selection objective, we should take the allele with the largest negative effect per main-effect QTL as novel allele. The same is true for allele combination of epistatic QTL. The summary statistics for novel allele or allele combination are given in Table 2. These results show that there is one novel allele for each main-effect locus or one novel allele combination for each epistatic QTL. For example, for the locus linked to marker satt656, all the allelic effects are showed in **Fig. 3**, and novel allele is the allele with an effect of 2.63. Similarly, for the interaction between markers sat_342 and AW277661, novel allele combination is the allele combination with an effect of 1.29. The novel allele and allele combination were found in the Zhengzhou 790034 and Guangxibayuehuang cultivars, respectively.

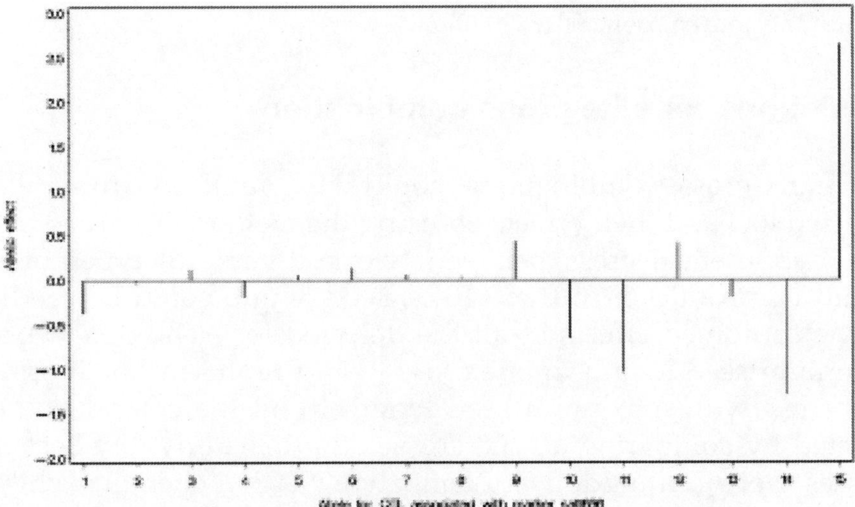

Figure 3. Allelic effects for QTL associated with marker satt656 for soybean seed length (mm).

doi:10.1371/journal.pone.0017773.g003

Table 2. The information of novel allele for QTL with r^2 larger than 5%.

QTL	Chr.	Marker associated	Position (cM)	Novel allele (bp)	Effect (mm)	Cultivar with novel allele
Main-effect	A1	satt382	26.42	295	0.64	Qinyan 1
	B2	satt534	87.59	185	1.22	Zhenghezhibanzi
	C2	sat_252	127.00	276	1.00	Taixinghanludou
	D2	satt514	85.69	242	1.11	Caishengzi
	D2	sat_365	87.39	286	0.95	Dandou 2
	F	satt656	135.12	182 or 170	2.63	Zhengzhou 790034
	G	satt352	50.53	178	0.87	Ya'anguanhualiyuebao
	K	satt441	46.20	282	1.11	Nannongdahuangdou
	N	satt022	102.06	277	0.94	Dandongdaliqing
Epsitasis	B2 & C1	sat_342 & AW277661	20.31 & 74.79	288 & 301	1.29	Guangxibayuehuang
	D1a & E	sat_160 & satt411	104.28 & 12.92	190 & 109	0.99	Anbaishuidou
	D1b & E	satt459 & satt411	118.62 & 12.92	195 or 189 & 106	1.09	Zhengzhou 74064

doi:10.1371/journal.pone.0017773.t002

doi:10.1371/journal.pone.0017773.t002

Predictions for elite cross combination

The elite cross combinations could be predicted from all the detected loci and their effects by using the method described below. In a hypothetical cross between two cultivars, all types of RILs would be produced. In these RILs, seed length could be predicted by the combined effects of all the detected loci. The best RIL with maximum seed length in one cross would represent the cross. The best cross with maximum seed length in all the crosses could be selected by comparing all the crosses. In this study, the best three crosses were Daqingdou × Zhengzhou790034, Zhenghe- zhibanzi × Zhengzhou790034, and Liyangdawuhuangdou × Zhengzhou 790034. The presence of Zhengzhou790034 in the three best crosses indicated that it contained the best allele or allele combination.

Monte Carlo simulation studies

Evaluation of the performance of the proposed approach.

The first simulation experiment was designed to investigate the effect of QTL heritability on QTL mapping in crop cultivars. The results show that the precision and power of the detection of QTLs increase with increasing QTL heritability, and that the false positive rate (FPR) is only 0.0244% (Table S2).

In the second simulation experiment, we investigated the effect of sample size by randomly sampling 100, 200, or 300 non-founder lines. The other parameters were the same as those in the first simulation experiment. As expected, the precision and power increased with increasing sample size (Table S3). Sample sizes under 300 yield much better results than those under 200; we recommend a sample size of 300 for future studies.

The third simulation experiment compared the effect of the number of alleles on QTL mapping in crop cultivars. We set the numbers of alleles at 2, 3 and 4; other parameters were the same as those in the first simulation experiment. The results showed that precision and power decrease as the number of alleles increases (Table S4). The results also imply that the SNP or indel markers are better than the other markers.

In the fourth simulation experiment, the effect of allelic frequency on QTL mapping was assessed by setting the frequency ratio of the two alleles as 1:1 (uniform distribution), 1:2 (skewed distribution) or 1:3 (skewed distribution). The other parameters were the same as those in the first simulation experiment. The results showed that skewed distribution decreased the statistical power (Table S5), indicating that rare alleles should be preferentially studied in association analyses.

The detection of QTL-by-environment interaction.

To investigate whether environmental effects could be detected, all the cultivars were evaluated in multiple environments. In the fifth simulation experiment, two environments, ten main-effect QTL and five QTL-by-environment interactions were simulated. The new method holds greater power for detecting QTL-by-environment interactions than for the main-effect QTL, and the FPR is lower than 0.06% (Table 3). To further demonstrate the performance of the new method, in the sixth simulation experiment, we designed a large genome with high density markers. In total, 510 markers were simulated on ten chromosome segments 1,000 cM long, with an average marker interval of 2 cM. The other parameters were the same as those in the fifth simulation experiment. The same trend in the fifth experiment was obtained (Table 4), indicating that our method works in large genomes with a high marker density.

Table 3. Environmental interaction detection in Monte Carlo simulation experiment (200 replicates).

doi:10.1371/journal.pone.0017773.t003

QTL	True value				Estimate			
	Chr.	Position (cM)	Variance	r^2 (%)	Power (%)	Position (cM)	Variance	r^2 (%)
Main-effect	1	70.3	0.926	5.0	100.0	70.3(0.0)	0.8934(0.2176)	4.94(1.21)
		262.8	0.926	5.0	99.5	262.8(0.0)	0.8912(0.2131)	4.92(1.15)
	2	401.4	0.370	2.0	95.0	401.4(0.0)	0.3552(0.1366)	1.96(0.74)
		438.8	0.556	3.0	99.0	438.8(0.0)	0.5215(0.1589)	2.88(0.86)
	3	601.6	0.926	5.0	100.0	601.6(0.0)	0.8816(0.2125)	4.87(1.15)
	8	1653.4	0.185	1.0	58.0	1653.4(0.4)	0.2097(0.0858)	1.15(0.47)
		1747.6	0.370	2.0	93.5	1747.6(0.0)	0.3384(0.1372)	1.87(0.76)
	9	1944.7	1.852	10.0	100.0	1944.7(0.0)	1.8511(0.3121)	10.22(1.59)
	10	2145.2	0.926	5.0	100.0	2145.2(0.0)	0.9322(0.2352)	5.15(1.23)
		2181.6	0.926	5.0	100.0	2181.6(0.0)	0.9081(0.2051)	5.02(1.09)
Environment			0.926	5.0	96.0		0.8744(0.2580)	4.82(1.39)
Environmental	1	55.6	0.463	2.5	97.0	55.6(0.0)	0.4229(0.1391)	2.33(0.75)
interaction	2	401.4	0.463	2.5	98.0	401.4(0.0)	0.4465(0.1678)	2.46(0.88)
		438.8	0.926	5.0	100.0	438.8(0.0)	0.8867(0.2100)	4.90 (1.12)
	3	682.7	0.926	5.0	100.0	682.7(0.0)	0.9016(0.2190)	4.98(1.19)
	8	1747.6	1.852	10.0	100.0	1747.6(0.0)	1.8344(0.2903)	10.13(1.47)
False positive rate (%)		0.0550						

doi:10.1371/journal.pone.0017773.t003

Table 4. Environmental interaction detection under the situations of large genome and high-density markers (200 replicates).

doi:10.1371/journal.pone.0017773.t004

The identification of QTL-by-QTL interaction.

To demonstrate whether QTL-by- QTL interactions could be detected, all epistatic effects between two main-effect QTLs were included in the full model. In the final simulation experiment, 50 markers were evenly distributed in five linkage groups 450 cM in length. Five main-effect QTLs, 3 QTL-by-environment interactions and 5 QTL-by-QTL interactions were simulated. The results (Table 5) show that the estimates for the positions and variances of simulated QTLs are close to their true values, and the power in the detection of

QTL is high (e.g., over 80% for the QTLs with a heritability over 2%), especially for QTL-by-QTL interactions.

QTL	True value				Estimate			
	Chr.	Position (cM)	Variance	r^2 (%)	Power (%)	Position (cM)	Variance	r^2 (%)
Main-effect	1	50	0.4	2	83.5	50.0(0.0)	0.3967(0.1317)	2.04(0.66)
	2	100	1.0	5	97.5	100.0(0.0)	0.9441(0.2544)	4.88(1.31)
	3	200	3.0	16	99.5	200.0(0.0)	1.8338(0.5818)	9.99(2.35)
	4	350	0.4	2	82.3	350.0(0.0)	0.3953(0.1371)	2.03(0.70)
	5	400	1.0	5	95.5	400.0(0.0)	[illegible]	4.58(1.21)
Environment			1.0	5	99.3		0.9438(0.2299)	4.88(1.14)
[illegible]	3	150	0.4	2	98.5	150.0(0.0)	0.3984(0.1231)	1.99(0.67)
Interaction	3	270	3.0	18	100.0	270.0(0.0)	1.8930(0.3807)	10.21(1.87)
	5	400	1.0	5	99.5	400.0(0.0)	0.9154(0.2361)	4.60(1.12)
Epistasis	1 & 2	10 & 110	0.4	2	97.0	10.0(3.4) & 120.9(5.4)	0.3444(0.1262)	1.78(0.65)
	2 & 3	100 & 250	1.0	5	100.0	100.0(0.0) & 250.0(0.0)	0.9625(0.2196)	5.09(1.13)
	3 & 5	200 & 400	0.4	2	85.5	200.0(0.0) & 395.9(1.5)	0.3642(0.1275)	1.98(0.66)
	3 & 4	270 & 350	3.0	18	100.0	270.1(0.7) & 350.0(1.0)	1.8358(0.3605)	9.29(1.79)
	4 & 5	350 & 450	3.0	18	100.0	350.1(0.7) & 450.0(0.0)	1.8614(0.3812)	10.25(1.80)
False positive rate (%)			0.0545					

doi:10.1371/journal.pone.0017773.t005

Table 5. Epistatic QTL detection in Monte Carlo simulation experiment (200 replicates).

doi:10.1371/journal.pone.0017773.t005

Discussion

The approach proposed in this work has several advantages over the approaches of previous association analysis studies. First, main, environmental, QTL-by- environment and QTL-by-QTL interactions were simultaneously considered in our full genetic model, improving the statistical power [38]–[41]. Although multi-locus genetic models have been proposed in plant genetics [59]–[62], they have difficulty combining both QTL-by-environment and QTL-by-QTL interactions. Epistasis association mapping has been developed in human genetics [33]–[37], but here the epistasis was identified by two-dimensional scan, and significant effects in the two-dimensional scan were further tested in one genetic model. Second, epistasis association analysis was first integrated with crop breeding by design. In the past, the results from QTL mapping have had limited utility in breeding practice, due to the use of a simple cross population or the neglect of epistasis in the detection of QTLs. We designed an elite cross combination to take these two issues into account. Third, it is easy to extend the proposed approach to nested association analysis. The commonality is that all the individuals in the mapping populations are inbred lines. The difference is that the pedigree is general for the present study and relatively simple for nested association analysis. Therefore, the new method is suitable for nested association analysis and human genetics. Fourth, the FPR is minimized in the new method. A shrinkage estimation method, empirical Bayes (eBayes), was adopted to estimate all types of effects in the full model so that the FPR was less than 0.06%.

At present the most widely used genome-wide association study (GWAS) is analysis of variance or mixed model approaches with the control of false discovery rate. In theory, it is similar to single-marker analysis for main-effect QTL and two-marker analysis for epistatic QTL, and the difference is that the GWAS requires the setting of a significance threshold at the genome-wide level. However, it does not overcome the shortcomings of marker analysis. If a trait of interest is controlled by multiple QTLs, whether the QTL under consideration can be detected depends on the proportions of phenotypic variance explained by both this QTL and background QTLs. If the proportion by background QTLs is large, large residual variance will result in a decreased power in the detection of the current QTL and sometime the QTL can not be identified. In the new approach, this issue can be avoided, because a full model that includes all kinds of QTL in one genetic model results in a small residual variance. This explains why some main-effect QTLs and all the epistatic QTLs can not be mapped in the soybean genome-wide association study.

Prediction of elite cross combination is based on the assumption that dominance and dominance-type epistasis effects are absent. If the breeding objective is the development of inbred lines or cultivars as often the case in self-pollinated crops, the prediction may be useful. If these non-additive effects are important, then the prediction would not reliable. This issue needs to be addressed in the future.

Xu [41] described a linear model in which the dimensions of the genotypic value vector and its incidence matrix depend on the number of genotypes for the locus. In theory, this model matches the situation under study. However, the model dimensions will increase rapidly. Therefore, it is preferable to gather more samples or reduce the number of effects considered[38], [63] to reduce the dimensions of the model. In this study, we designed a special incidence matrix such that there is one variable for each main-effect QTL. Simulation studies show that this approach works well. If the number of markers is large, the number of effects in the model is enormous. In this case, the two-stage method of He and Zhang [58] is recommended. We adopted this approach in our analysis of real data, and the results were consistent with those of He and Zhang [58] and He et al. [64]. The new approach works well if the marker interval length is approximately 5 cM. However, one must delete some closely linked markers if the interval length is less than 5 cM [64].

We compared the QTLs of seed length in soybeans with the QTLs in previous studies. Although few common markers existed between their data and ours, some loci that we detected were also detected in previous studies. Seven QTLs linked to markers sat_342, satt534, satt514, sat_365, sat_254, sat_419 and sat_274 in this study were detected by Xu et al. [65]; four QTLs associated with markers satt411, satt329, satt022 and AW277661 in this paper were identified by Salas et al. [66]; one QTL close to marker sat_256 was confirmed by Li et al. [67]; and one QTL next to marker satt514 was mapped by Liang et al.[68]. The above results further confirmed the feasibility of the approach proposed in this study.

MATERIALS AND METHODS

Soybean samples

We recently assembled a soybean association panel with 215 cultivars provided by the National Center for Soybean Improvement, China. All the cultivars were obtained by stratified random sampling from six geographic ecotypes in China [69], planted in three-row plots in a completely randomized design and evaluated at the Jiangpu experimental station at Nanjing Agricultural University in 2008 and 2009. The plots were 1.5 m wide and 2 m long. Five individuals and 20 seeds in the middle row of each plot were randomly picked to measure seed length by digital vernier caliper. The measurements were averaged over 20 seeds, and the mean was used in this study.

Approximately 0.3 g of fresh leaves obtained in 2008 from each cultivar was used to extract genomic DNA using the cetyltrimethylammonium bromide method as described by Lipp et al.[70]. To screen for polymorphisms among all the cultivars, PCR was performed with 134 simple sequence repeat (SSR) primer pairs. The primer sequences were obtained from the soybean database Soybase (http://www.ncbi.nlm.nih.gov). PCR was performed as described by Xu et al. [65].

Population structure

For the soybean data, the STRUCTURE program was used to investigate the population structures of all selected cultivars [26]. The number of subpopulations (*K*) was set from 2 to 10. In the Markov chain Monte Carlo (MCMC) Bayesian analysis for each *K*, the length of a Markov chain consisted of 110,000 sweeps. The first 10,000 sweeps (the burn-in period) were deleted, and thereafter, the chain was used to calculate the mean of log-likelihood. This process was repeated 20 times, and the total average for mean log-likelihood at fixed *K* was used. STRUCTURE analysis with 134 SSR molecular markers showed that the log-likelihood increased with the increase of the model parameter *K*, so a suitable number of *K* could not be determined. In this situation, using the ad hoc statistic ΔK, based on the rate of change in the log-probability of data between successive *K* values, STRUCTURE accurately detected the uppermost hierarchical level of structure [71]. Here, the ΔK value was much higher for the model parameter $K=4$ than for other values of *K*. By combining this high ΔK value with knowledge of the breeding history of these cultivars, we chose a value of 4 for *K*. The Q matrix was calculated based on SSR markers and incorporated into the mixed model of epistasis association analysis.

Genetic model

The phenotypic value of a quantitative trait for the *i*th cultivar in the *j*th environment ($i=1,\cdots,n; j=1,\cdots,R$), y_{ij}, may be described by the following mixed model:

$$\mathbf{Y}=\mu+\sum_{l=1}^{K-1}\mathbf{X}_{\mathbf{P}l}\beta_{\mathbf{P}l}+\mathbf{X}_{\mathbf{E}}\beta_{\mathbf{E}}+\sum_{s=1}^{m}\mathbf{Z}_{\mathbf{Q}s}\gamma_{\mathbf{Q}s}+\sum_{s=1}^{m}\mathbf{Z}_{\mathbf{QE}s}\gamma_{\mathbf{QE}s}+\sum_{s=1}^{m-1}\sum_{t=s+1}^{m}\mathbf{Z}_{\mathbf{QQ}st}\gamma_{\mathbf{QQ}st}+\varepsilon \quad (1)$$

where $\mathbf{Y}=(y_{11},\cdots,y_{1n},\cdots,y_{R1},\cdots,y_{Rn})$; $\mathbf{X}_{\mathbf{P}}=(\mathbf{X}_{\mathbf{P}1},\cdots,\mathbf{X}_{\mathbf{P},K-1})$ is the Q matrix for population structure; $\mathbf{X}_{\mathbf{E}},\mathbf{Z}_{\mathbf{Q}}=(\mathbf{Z}_{\mathbf{Q}1},\cdots,\mathbf{Z}_{\mathbf{Q}m}),\mathbf{Z}_{\mathbf{QE}}=(\mathbf{Z}_{\mathbf{QE}1},\cdots,\mathbf{Z}_{\mathbf{QE}m})$ and

$\mathbf{Z_{QQ}} = (\mathbf{Z_{QQ11}}, \cdots, \mathbf{Z_{QQ(m-1)m}})$ are the design matrices of the environment effect, main effect, QTL-by-environment interaction effect and QTL-by-QTL interaction effect, respectively; $\boldsymbol{\beta}_\mathbf{P} = (\boldsymbol{\beta}_\mathbf{P1}, \cdots, \boldsymbol{\beta}_{\mathbf{P},K-1})', \boldsymbol{\beta}_\mathbf{E}, \boldsymbol{\gamma}_\mathbf{Q} = (\boldsymbol{\gamma}_\mathbf{Q1}, \cdots, \boldsymbol{\gamma}_{\mathbf{Q}m})', \boldsymbol{\gamma}_\mathbf{QE} = (\boldsymbol{\gamma}_\mathbf{QE1}, \cdots, \boldsymbol{\gamma}_{\mathbf{QE}m})'$ and $\boldsymbol{\gamma}_\mathbf{QQ} = (\boldsymbol{\gamma}_\mathbf{QQ11}, \cdots, \boldsymbol{\gamma}_{\mathbf{QQ},(m-1)m})'$ are the corresponding effects; and $\boldsymbol{\mu}$ is the total average. The first three terms were viewed as fixed effects and the following three terms were considered random effects; therefore, model (1) was rewritten as $\mathbf{Y} = \mathbf{X}\boldsymbol{\beta} + \mathbf{Z}\boldsymbol{\gamma} + \boldsymbol{\varepsilon}$ **(2)** where $\mathbf{X} = (\mathbf{1}\,\mathbf{X_P}\,\mathbf{X_E})$, $\mathbf{Z} = (\mathbf{Z_Q}\,\mathbf{Z_{QE}}\,\mathbf{Z_{QQ}})$, $\boldsymbol{\beta} = (\boldsymbol{\mu}, \boldsymbol{\beta}'_\mathbf{P}, \boldsymbol{\beta}'_\mathbf{E})'$ and $\boldsymbol{\gamma} = (\boldsymbol{\gamma}'_\mathbf{Q}, \boldsymbol{\gamma}'_\mathbf{QE}, \boldsymbol{\gamma}'_\mathbf{QQ})'$.

Parameter estimation

Several methods exist to simultaneously estimate the parameters in model (2); for example, eBayes [41], [43]. Here, we adopted eBayes. Briefly, the parameter vector in model (2) is $\boldsymbol{\theta} = (\boldsymbol{\beta}\boldsymbol{\gamma}\boldsymbol{\sigma}^2)$. The priors and the likelihood are not described in detail here. The iteration process is given below.

The fixed effects were calculated by:

$$\boldsymbol{\beta}^{(t+1)} = [\mathbf{X}^T(\mathbf{V}^{(t)})^{-1}\mathbf{X}]^{-1}\mathbf{X}^T(\mathbf{V}^{(t)})^{-1}\mathbf{Y} \quad \textbf{(3)}$$

$$\boldsymbol{\sigma}^{2(t+1)} = \frac{\boldsymbol{\sigma}^{2(t)}}{n}(\mathbf{Y} - \mathbf{X}\boldsymbol{\beta}^{(t)})^T(\mathbf{V}^{(t)})^{-1}(\mathbf{Y} - \mathbf{X}\boldsymbol{\beta}^{(t)}) \quad \textbf{(4)}$$

where $\mathbf{V} = \sum_{j=1}^{m} \mathbf{Z}_j\mathbf{Z}_j^T\boldsymbol{\sigma}_j^2 + \mathbf{I}\boldsymbol{\sigma}^2$. Note that there is not an explicit solution for the estimation of $\boldsymbol{\sigma}_j^2$, and it is updated by maximizing

$$L(\boldsymbol{\sigma}_j^2|\cdots) = -\frac{1}{2}\ln(\mathbf{Z}_j^T(\mathbf{V}^{(t)})^{-1}\mathbf{Z}_j(\boldsymbol{\sigma}_j^2 - \boldsymbol{\sigma}_j^{2(t)}) + 1) +$$

$$\frac{(\boldsymbol{\sigma}_j^2 - \boldsymbol{\sigma}_j^{2(t)})[(\mathbf{Y} - \mathbf{X}\boldsymbol{\beta}^{(t)})^T(\mathbf{V}^{(t)})^{-1}\mathbf{Z}_j]^2}{2(\mathbf{Z}_j^T(\mathbf{V}^{(t)})^{-1}\mathbf{Z}_j)(\boldsymbol{\sigma}_j^2 - \boldsymbol{\sigma}_j^{2(t)}) + 1} - \frac{1}{2}(\tau + 2)\ln\boldsymbol{\sigma}_j^2 - \frac{\omega}{2\boldsymbol{\sigma}_j^2} \quad \textbf{(5)}$$

where $\tau = -1.0$ and $\omega = 0.0005$.

The random effects, $\boldsymbol{\gamma}_j$, were predicted by best linear unbiased prediction (BLUP): $E(\boldsymbol{\gamma}_j|\mathbf{Y}) = \boldsymbol{\sigma}_j^2\mathbf{Z}_j^T\mathbf{V}^{-1}(\mathbf{Y} - \mathbf{X}\boldsymbol{\beta})$ **(6)**

The posterior variance of $\boldsymbol{\gamma}_j$ is $\mathrm{var}(\boldsymbol{\gamma}_j|\mathbf{Y}) = \boldsymbol{\sigma}_j^2(1 - \mathbf{Z}_j^T\mathbf{V}^{-1}\mathbf{Z}_j\boldsymbol{\sigma}_j^2)$ **(7)**

The proportion of phenotype variance explained by one random effect may be calculated by $h_j^2 \approx \sigma_{Z_j}^2 \gamma_j^2 / \sigma_P^2$ (8)

Likelihood ratio test

The traditional likelihood ratio test (LRT), as described by Zhang and Xu [38], could not be performed in this study, due to an oversaturated epistatic genetic model. We proposed the following two-stage selection process to screen all the effects. In the first stage, all the effects with $|\gamma_j / \sigma| > 10^{-6}$ are picked up. In the second stage, the full model is modified so that only the effects that passed the first round of selection are included. Due to the smaller dimensionality of the reduced model, we can use the maximum likelihood method to reanalyze the data and perform the LRT. The procedure for the LRT is below.

The overall null hypothesis is no effect of the QTL at the locus of interest, denoted by $H_0 : a_1 = \cdots = a_T = 0$, where a_t is the effect of the tth allele. If we solve the maximum likelihood estimation of the parameters under the restriction of $H_0 : a_1 = \cdots = a_T = 0$ and calculate the log-likelihood value using the solutions with this restriction, we obtain $L(\hat{\theta}|H_0)$. We can also evaluate the log-likelihood value of the solutions without restrictions and obtain $L(\hat{\theta})$. Therefore, the LR test statistic $\mathbf{LR} = -2\left[L(\hat{\theta}|\mathbf{Lu}=0) - L(\hat{\theta})\right]$. (9)

Other test statistics can be used in similar ways. The significance threshold of the LOD score was set at 2.5 for our real data analysis, where $\mathrm{LOD} = \mathrm{LR}/4.605$.

Genome-wide association study

First, phenotypic values for seed length in 215 soybean cultivars were corrected using population structure obtained by STRUCTURE software. Then, the corrected phenotypes along with SSR marker information were used to carry out genome-wide association studies for main-effect QTLs, environmental interactions and QTL-by- QTL interactions by ANOVA. Finally, critical values at the 0.05 level of significance were determined by 1000 permutation experiments and thus significant QTL could be identified.

SIMULATION DESIGN

We performed seven simulation experiments in this study. In the first, the simulated pedigree was the maize pedigree described by Zhang et al. [31], [61]. The number of inbred lines within the maize pedigree was 404(n). Of these, $n_0(=103)$ were base (founder) lines, which were in linkage equilibrium so that the genotypes for markers and QTLs with two alleles could be simulated. Non-founders (n_1 = 301) were bred via repeated self-pollination of a hybrid between two inbred lines. Thus, each non-founder line represents a recombinant inbred line (RIL) with respect to a known pair of parents. The genotypes of all the non-founders could be generated from the genotypes of their parents, analogous to simulating the genotypes of RILs from their parents. All of the non-founder lines could be used to detect QTLs. To mimic the actual linkage maps that did not have equally spaced markers, 153 markers were simulated on ten chromosome segments of length ~2258.70 cM, with an average marker interval of 14.86 cM. A total of 20 QTLs, all of which overlapped with the markers, were simulated; the sizes and locations of the QTLs are listed in Table 3. The allelic effects were calculated by relating the genetic variance of the QTL to both the allelic frequencies and the allelic number. The phenotypic value of each line was the sum of the corresponding QTL genotypic values and the residual error, with an assumed normal distribution. Each simulation run consisted of 200 replicates. For each simulated QTL, we counted the samples in which the LOD statistic surpassed 3.0. The ratio of the number of such samples (m) to the total number of replicates (200) represented the empirical power of this QTL. The false-positive rate was calculated as the ratio of the number of false-positive effects to the total number of zero effects considered in the full model. The other simulation experiments were performed similarly. All simulated parameters are given in Table S1.

ACKNOWLEDGMENTS

We are grateful to three anonymous referees for their constructive comments and suggestions that significantly improved the presentation of the manuscript.

AUTHOR CONTRIBUTIONS

Conceived and designed the experiments: YMZ. Performed the experiments: XFL HYL SPW. Analyzed the data: HYL. Contributed reagents/materials/analysis tools: HYL XFL SPW. Wrote the paper: YMZ HYL.

REFERENCES

1. Abdalla AM, Reddy OUK, El-Zik KM, Pepper AE (2001) Genetic diversity and relationships of diploid and tetraploid cottons revealed using AFLP. Theor Appl Genet 102: 222–229.
2. Dong YS, Zhuang BC, Zhao LM, Sun H, He MY (2001) The genetic diversity of annual wild soybeans grown in China. Theor Appl Genet 103: 98–103.
3. Reif JC, Hamrit S, Heckenberger M, Schipprack W, Maurer HP, et al. (2005) Genetic structure and diversity of European flint maize populations determined with SSR analyses of individuals and bulks. Theor Appl Genet 111: 906–913.
4. Milne RI, Abbott RJ (2000) Origin and evolution of invasive naturalized material of*Rhododendron ponticum* L. in the British isles. Mol Ecol 9(5): 541–556.
5. Dillon SL, Shapter FM, Henry RJ, Cordeiro G, Liz Izquierdo, Liz LS (2007) Domestication to crop improvement: Genetic resources for Sorghum and saccharum (Andropogoneae). Annals of Botany 100: 975–989.
6. Friesen ML, von Wettberg EJ (2010) Adapting genomics to study the evolution and ecology of agricultural systems. Current Opinion in Plant Biology 13: 119–125.
7. Ellis RP, Forster BP, Robinson D, Handley LL, Gordon DC, et al. (2000) Wild barley: a source of genes for crop improvement in the 21 century? J Exp Bot 51: 9–17.
8. Upadhyaya HD, Ortiz R (2001) A mini core subset for capturing diversity and promoting utilization of chickpea genetic resources in crop improvement. Theor Appl Genet 102: 1292–1298.
9. Warburton ML, Crossa J, Franco J, Kazi M, Trethowan R, et al. (2006) Bringing wild relatives back into the family: recovering genetic diversity in CIMMYT improved wheat germplasm. Euphytica 149: 289–301.

10. Sasaki A, Ashikari M, Ueguchi-Tanaka M, Itoh H, Nishimura A, et al. (2002) Green revolution: a mutant gibberellin-synthesis gene in rice. Nature 416(6882): 701–702.
11. Zhang X-Q (2002) Three lines hybrid rice. In: Shi Y-C, editor. Chinese Academic canon in the 20th century. Fuzhou: Fujian Education Press. pp. 25–27.
12. Stuber CW (1995) Mapping and manipulating quantitative trait in maize. Trends in Genetics 11: 477–481.
13. Tanksley SD, Nelson JC (1996) Advanced backcross QTL analysis: a method for the simultaneous discovery and transfer of valuable QTLs from unadapted germplasm into elite breeding lines. Theor Appl Genet 92: 191–203.
14. Haseman JK, Elston RC (1972) The investigation of linkage between a quantitative trait and a marker locus. Behav Genet 2: 3–19.
15. Wright FA (1997) The phenotypic difference discards sib-pair QTL linkage information. Am J Hum Genet 60: 740–742.
16. Drigalenko E (1998) How sib pairs reveal linkage. Am J Hum Genet 63: 1242–1245.
17. Forrest W (2001) Weighting improves the "new Haseman-Elston" method. Hum Hered 52: 47–54.
18. Sham PC, Purcell S (2001) Equivalence between Haseman-Elston and variance components linkage analyses for sib pairs. Am J Hum Genet 68: 1527–1532.
19. Sham PC, Purcell S, Cherny SS, Abecasis GR (2002) Powerful regression -based quantitative trait linkage analysis of general pedigrees. Am J Hum Genet 71(2): 238–253.
20. Chen WM, Broman KW, Liang KY (2004) Quantitative trait linkage analysis by generalized estimating equations: unification of variance components and Haseman-Elston regression. Genet Epidemiol 26(4): 265–272.
21. Wang T, Elston RC (2005) Two-level Haseman-Elston regression for general pedigree data analysis. Genet Epidemiol 29(1): 12–22.
22. Sax K (1923) The association of size difference with seed-coat pattern and pigmentation in *Phaseolus vulgaris*. Genetics 8: 552–560.
23. Risch N, Merikangas K (1996) The future of genetic studies of complex human diseases. Science 273: 1516–1517.
24. Smouse PE, Waples RS, Tworek JA (1990) A genetic mixture analysis for use with incomplete source population-data. Can J Fish Aquat Sci 47: 620–634.

25. Balding DJ, Nichols RA (1995) A method for quantifying differentiation between populations at multi-allelic loci and its implications for investigating identity and paternity. Genetica 96: 3–12.
26. Pritchard JK, Stephens M, Donnelly P (2000) Inference of population structure using multilocus genotype data. Genetics 155: 945–959.
27. Marchini J, Cardon LR, Phillips MS, Donnelly P (2004) The effects of human population structure on large genetic association studies. Nat Genet 36: 512–517.
28. Zhu CS, Yu JM (2009) Nonmetric multidimensional scaling corrects for population structure in association mapping with different sample types. Genetics 182: 875–888.
29. Li MY, Reilly MP, Rader DJ, Wang LS (2010) Correcting population stratification in genetic association studies using a phylogenetic approach. Bioinformatics 26(6): 798–806.
30. Diao G, Lin DY (2005) A powerful and robust method for mapping quantitative trait loci in general pedigrees. Am J Hum Genet 77: 97–111.
31. Zhang Y-M, Mao YC, Xie C, Smith H, Luo L, et al. (2005) Mapping QTL using naturally occurring genetic variance among commercial inbred lines of maize (*Zea mays* L.). Genetics 169: 2267–2275.
32. Yu J, Pressoir G, Briggs WH, Vroh Bi I, Yamasaki M, et al. (2006) A unified mixed-model method for association mapping that accounts for multiple levels of relatedness. Nat Genet 38(2): 203–208.
33. Chatterjee N, Kalaylioglu Z, Moslehi R, Peters U, Wacholder S (2006) Powerful multilocus tests of genetic association in the presence of gene-gene and gene-environment interactions. Am J Hum Genet 79(6): 1002–1016.
34. Chen X, Liu CT, Zhang MZ, Zhang HP (2007) A forest-based approach to identifying gene and gene-gene interactions. Proc Natl Acad Sci USA 104: 19199–19203.
35. Zhang Y, Liu JS (2007) Bayesian inference of epistatic interactions in case-control studies. Nat Genet 39: 1167–1173.
36. Phillips P (2008) Epistasis — the essential role of gene interactions in the structure and evolution of genetic systems. Nat Rev Genet 9: 855–867.
37. Wan X, Yang C, Yang Q, Xue H, Tang NLS, et al. (2009) MegaSNPHunter: a learning approach to detect disease predisposition SNPs and high level interactions in genome wide association study. BMC Bioinformatics 10: 13.

38. Zhang Y-M, Xu S (2005) A penalized maximum likelihood method for estimating epistatic effects of QTL. Heredity 95: 96–104.
39. Xu S, Jia Z (2007) Genome-wide analysis of epistatic effects for quantitative traits in Barley. Genetics 175: 1955–1963.
40. Hoggart CJ, Whittaker JC, De Iorio M, Balding DJ (2008) Simultaneous analysis of all SNPs in genome-wide and re-sequencing association studies. PLoS Genet 4(7): e1000130.
41. Xu S (2010) An expectation–maximization algorithm for the Lasso estimation of quantitative trait locus effects. Heredity 105: 483–494.
42. Tibshirani R (1996) Regression shrinkage and selection via the lasso. Journal of the Royal Statistical Society Series B 58: 267–288.
43. Xu S (2007) An empirical Bayes method for estimating epistatic effects of quantitative trait loci. Biometrics 63: 513–521.
44. Bernardo R (2008) Molecular markers and selection for complex traits in plants: learning from the last 20 years. Crop Sci 48: 1649–1664.
45. Holland JB (2004) Implementation of molecular markers for quantitative traits in breeding programs - challenges and opportunities. In New directions for a diverse planet, Proceedings of the 4th International Crop Science Congress, 26 Sep – 1 Oct 2004, Brisbane, Australia. Published on CDROM. Web sitehttp://www.cropscience.org.au/.
46. Michelmore RW, Paran I, Kesseli RV (1991) Identification of markers linked to disease resistance genes by bulked-segregant analysis: a rapid method to detect markers in specific genomic regions by using segregating populations. Proc Natl Acad Sci USA 88: 9828–9832.
47. Cho YG, Eun MY, McCouch SR, Chae YA (1994) The semidwarf gene, *sd-1*, of rice (*Oryza sativa* L.).II. Molecular mapping and marker-assisted selection. Theor Appl Genet 89: 54–59.
48. Mohan M, Nair S, Bhagwat A, Krishna TG, Yano M, et al. (1997) Genome mapping, molecular markers and marker-assisted selection in crop plants. Molecular Breeding 3: 87–103.
49. Ribaut JM, Betrán J (1999) Single large-scale marker-assisted selection (SLS-MAS). Molecular Breeding 5: 531–541.
50. Zhang TZ, Yuan Y, Yu J, Guo WZ, Kohel RJ (2003) Molecular tagging of a major QTL for fiber strength in upland cotton and its marker-assisted selection. Theor Appl Genet 106: 262–268.
51. Jahufer MZZ, Cooper M, Ayres JF, Bray RA (2002) Identification of research to improve the efficiency of breeding strategies for white clover in Australia: A review. Australian Journal of Agricultural Research 53(3): 239–257.

52. Dwivedi SL, Crouch JH, Mackill DJ, Xu YB, Blair MW, et al. (2007) The molecularization of public sector crop breeding: Progress, problems, and prospects. Advances in Agronomy 95: 163–318.
53. Carlborg Ö, Haley CS (2004) Epistasis: too often neglected in complex trait studies? Nat Rev Genet 5: 618–625.
54. Peleman JD, van der Voort JR (2003) Breeding by design. Trends in Plant Sci 8: 330–334.
55. Liu YF, Zeng ZB (2000) A general mixture model approach for mapping quantitative trait loci from diverse cross designs involving multiple inbred lines. Genet Res 75: 345–355.
56. Blanc G, Charcosset A, Mangin B, Gallais A, Moreau L (2006) Connected populations for detecting quantitative trait loci and testing for epistasis: an application in maize. Theor Appl Genet 113: 206–224.
57. Verhoeven KJF, Jannink JL, Mcintyre LM (2006) Using mating designs to uncover QTL and the genetic architecture of complex traits. Heredity 96: 139–149.
58. He XH, Zhang Y-M (2008) Mapping epistatic quantitative trait loci underlying endosperm traits using all markers on the entire genome in a random hybridization design. Heredity 101: 39–47.
59. Iwata H, Uga Y, Yoshioka Y, Ebana K, Hayashi T (2007) Bayesian association mapping of multiple quantitative trait loci and its application to the analysis of genetic variation among *Oryza sativa* L. germplasm. Theor Appl Genet 114: 1437–1449.
60. Iwata H, Ebana K, Fukuoka S, Jannink J-L, Hayashi T (2009) Bayesian multilocus association mapping on ordinal and censored traits and its application to the analysis of genetic variation among *Oryza sativa* L. germplasm. Theor Appl Genet 118: 865–880.
61. Zhang Y-M, Lü H-Y, Yao L-L (2008) Multiple quantitative trait loci Haseman-Elston regression using all markers on the entire genome. Theor Appl Genet 117: 683–690.
62. Lü H-Y, Li M, Li G-J, Yao L-L, Zhang Y-M (2009) Multiple loci *in silico* mapping in inbred lines. Heredity 103: 346–354.
63. Hoti F, Sillanpää MJ (2006) Bayesian mapping of genotype×expression interaction in quantitative and qualitative traits. Heredity 97: 4–18.
64. He X-H, Qin H, Hu Z, Zhang T, Zhang Y-M (2011) Mapping of epistatic quantitative trait loci in four-way crosses. Theor Appl Genet 122: 33–48.
65. Xu Y, Li HN, Li GJ, Wang X, Cheng LG, et al. (2011) Mapping quantitative trait loci for seed size traits in soybean (Glycine max L. Merr.). Theor Appl Genet 122: 581–594.

66. Salas P, Oyarzo-Llaipen JC, Wang D, Chase K, Mansur L (2006) Genetic mapping of seed shape in three populations of recombinant inbred lines of soybean (*Glycine max*L. Merr.). Theor Appl Genet 113: 1459–1466.
67. Li CD, Jiang HW, Zhang WB, Qiu PC, Liu CY, et al. (2008) QTL analysis of seed and pod traits in soybean. Molecular Plant Breeding 6: 1091–1100.
68. Liang HZ, Wang SF, Yu YL, Wang TF, Gong PT, et al. (2008) Mapping quantitative trait loci for six seed shape traits in soybean. Henan Agricultural Science 45: 54–60.
69. Wang YS, Gai JY (2002) Study on the ecological regions of soybean in China II. Ecological environment and representative varieties. Chinese Journal of Applied Ecology 13: 71–75.
70. Lipp M, Brodmann P, Pietsch K, Pauwels J, Anklam E, et al. (1999) IUPAC collaborative trail study of a method to detect genetically modified soybeans and maize in dried powder. Journal of AOAC International 82: 923–928.
71. Evanno G, Regnaut S, Goudet J (2005) Detecting the number of clusters of individuals using the software STRUCTURE: a simulation study. Molecular Ecology 14: 2611–2620.

Chapter 9

DEVELOPMENT AND EVALUATION OF A 9K SNP ARRAY FOR PEACH BY INTERNATIONALLY COORDINATED SNP DETECTION AND VALIDATION IN BREEDING GERMPLASM

[1]Ignazio Verde, [2]Nahla Bassil, [3]Simone Scalabrin, [2]Barbara Gilmore, [4]Cynthia T. Lawley, [5]Ksenija Gasic, [6]Diego Micheletti, [7]Umesh R. Rosyara, [3]Federica Cattonaro, [1]Elisa Vendramin, [8]Dorrie Main, [1]Valeria Aramini, [6]Andrea L. Blas, [9]Todd C. Mockler, [9,11]Douglas W. Bryant, [12]Larry Wilhelm, [13]Michela Troggio, [14]Bryon Sosinski, [6]Maria José Aranzana, [6]Pere Arús, [7]Amy Iezzoni, [3,15]Michele Morgante, [8]Cameron Peace

[1]Fruit Tree Research Center, Rome, Italy

[2]United States Department of Agriculture-Agricultural Research Service, National Clonal Germplasm Repository, Corvallis, Oregon, United States of America

[3]Istituto di Genomica Applicata, Udine, Italy

[4]Illumina Inc., Hayward, California, United States of America

[5]School of Agricultural, Forest and Environmental Sciences, Clemson University, Clemson, South Carolina, United States of America

[6]IRTA, Centre de Recerca en Agrigenòmica CSIC-IRTA-UAB-UB, Campus UAB, Cerdanyola del Vallès (Bellaterra), Barcelona, Spain

[7]Department of Horticulture, Michigan State University, East Lansing, Michigan, United States of America

[8]Department of Horticulture & Landscape Architecture, Washington State University, Pullman, Washington, United States of America

[9]The Donald Danforth Plant Science Center, St. Louis, Missouri, United States of America

[10]Department of Botany and Plant Pathology and Center for Genome Research and Biocomputing, Oregon State University, Corvallis, Oregon, United States of America

[11]Intuitive Genomics, Inc., St. Louis, Missouri, United States of America

[12]Oregon Health Sciences University, Portland, Oregon, United States of America

[13]The Istituto Agrario di San Michele all'Adige Research and Innovation Centre, Foundation Edmund Mach, San Michele all'Adige, Italy

[14]Department of Horticultural Science, North Carolina State University, Raleigh, North Carolina, United States of America

[15]Dipartimento di Scienze Agrarie e Ambientali, University of Udine, Udine, Italy

ABSTRACT

Although a large number of single nucleotide polymorphism (SNP) markers covering the entire genome are needed to enable molecular breeding efforts such as genome wide association studies, fine mapping, genomic selection and marker-assisted selection in peach [*Prunus persica* (L.) Batsch] and related *Prunus* species, only a limited number of genetic markers, including simple sequence repeats (SSRs), have been available to date.

To address this need, an international consortium (The International Peach SNP Consortium; IPSC) has pursued a coordinated effort to perform genome-scale SNP discovery in peach using next generation sequencing platforms to develop and characterize a high-throughput Illumina Infinium® SNP genotyping array platform. We performed whole genome re-sequencing of 56 peach breeding accessions using the Illumina and Roche/454 sequencing technologies. Polymorphism detection algorithms identified a total of 1,022,354 SNPs. Validation with the Illumina GoldenGate® assay was performed on a subset of the predicted SNPs, verifying ~75% of genic (exonic and intronic) SNPs, whereas only about a third of intergenic SNPs were verified. Conservative filtering was applied to arrive at a set of 8,144 SNPs that were included on the IPSC peach SNP array v1, distributed over all eight peach chromosomes with an average spacing of 26.7 kb between SNPs. Use of this platform to screen a total of 709 accessions of peach in two separate evaluation panels identified a total of 6,869 (84.3%) polymorphic SNPs.

The almost 7,000 SNPs verified as polymorphic through extensive empirical evaluation represent an excellent source of markers for future studies in genetic relatedness, genetic mapping, and dissecting the genetic architecture of complex agricultural traits. The IPSC peach SNP array v1 is commercially available and we expect that it will be used worldwide for genetic studies in peach and related stone fruit and nut species.

INTRODUCTION

Dissection of the genetic components underlying complex agricultural traits in plants has so far used mainly experimental bi-parental crosses and a limited number of genetic markers. The growing number of draft genome sequences of many species coupled with Next Generation Sequencing (NGS) technologies has quickly changed the paradigm of plant genome analysis. Alignment of the short NGS reads obtained from diverse and unrelated varieties to a reference genome allows the identification of large numbers of Single Nucleotide Polymorphisms (SNPs) and small Deletion/Insertion Polymorphisms (DIPs) and the contextual estimation of their minor allele frequencies (MAF).

Simultaneous genotyping of hundreds of thousands of SNPs in a single assay has become feasible due to innovative combinations of assay and array platform multiplexing [1]. Illumina's Infinium® BeadArray Technology platform is an extremely high-throughput

SNP genotyping system that allows the detection of up to 2.5 million SNPs per single DNA sample [2]. Multiplex SNP genotyping enables cost effective marker-assisted selection strategies, whole genome fingerprinting, genome-wide association studies (GWAS), map-based gene cloning and population-based analyses. The availability of such tools fosters the application of GWAS in plants and animals [3], [4], [5], [6] and provides the opportunity to apply genomic selection (GS) methods to agricultural species, including *Prunus*. High-density SNP genotyping arrays have been designed for several domestic animals including cattle [3], pig [4] and chicken [5]; arrays are being developed in several plant species including apple [6], maize, tomato, and cherry (http://www.illumina.com/agriculture).

In peach [*Prunus persica* L. (Batsch)] and related *Prunus* species, QTL studies have been conducted using experimental bi-parental crosses and a limited number of genetic markers [7],[8],[9],[10],[11],[12],[13]. A recent association study by Aranzana et al. [14] using a limited number (50) of SSR markers in unrelated accessions from American and European origin indicated that linkage disequilibrium (LD) in peach is quite high, up to 13–15 cM with stratification of peach accessions into at least three sub-populations. This study suggests that a small number of markers (about 600) might be sufficient to scan the peach genome. However, the peach gene pool used in this study [14] is known to have a narrow genetic base [15] in comparison with Eastern germplasm [16] where LD is likely to have a lower level of conservation. Moreover, studies in grape [17], [18] and maize [19] suggest that SNPs estimate a much lower decay of LD than SSRs. Hence, we suggest that a higher number of SNP markers covering the entire genome would be necessary to scan the peach genome and to perform GWAS.

The availability of the peach reference genome sequence recently released by the International Peach Genome Initiative [20] facilitated genome wide variant detection and the development of dedicated genomic tools. This and the ability to acquire massive sequence datasets from next generation sequencers, allowed the efficient identification of a large number of genetic markers, such as SNPs and small DIPs, enabling the development of a SNP array in this important horticultural crop. We describe here how members of The International Peach SNP Consortium (IPSC), that includes institutions from the U.S., Italy, and Spain, have worked together to identify genome-wide sequence variation and to develop a moderate-density peach high-throughput Infinium® SNP genotyping platform relevant for worldwide peach breeding germplasm, utilizing SNPs discovered using next generation sequencing platforms.

MATERIALS AND METHODS

Whole genome re-sequencing of peach breeding accessions

A SNP detection panel of 56 peach breeding-relevant accessions assembled with the goal of achieving an efficient coverage of the genetic background of cultivated peach (Table 1) was used for low-depth genome re-sequencing. The accessions were founders, intermediate ancestors, and important breeding parents used in international peach breeding programs, chosen both for the significance of their contribution to breeding germplasm according to pedigree records, and for genetic diversity based upon relatedness estimates from SSR studies[21], [14], (Verde I. unpublished data). Accessions were divided into 12 pools. For each accession in pools 1 through 11, paired-end libraries were prepared as recommended by Illumina (Illumina Inc., San Diego, CA, USA) separately at the USDA-ARS-National Clonal Germplasm Repository (NCGR) and the Istituto di Genomica Applicata (IGA, Udine, Italy) laboratories. In summary, library preparations were performed using minor modifications of the Illumina DNA-seq Sample Preparation protocol (Illumina, Inc., San Diego, CA). Briefly, 1–3 μg of genomic DNA was sheared by sonication using Diagenode's Biorupter XL sonicator system (Sparta, NJ, USA). This was followed by standard blunt-ending and 'A' was performed. Then, Illumina adapters with indexes (3 bp or 6 bp, at the NCGR and IGA, respectively) were ligated to the ends of the fragments. After the ligation reaction and separation of un-ligated adapters, samples were amplified by PCR to selectively enrich for those fragments in the library with adapter molecules at both ends. The samples were quantitated and quality tested using the NanoDrop ND-1000 UV-Vis Spectrophotometer (Thermo Scientific, Wilmington, DE, USA) and Agilent 2100 Bioanalyzer (Agilent Technologies, Santa Clara, CA). Libraries were pooled in equimolar ratios to yield a total concentration of 10 nM.

Table 1. Accessions of peach, almond and peach x almond hybrid sequenced at the Istituto di Genomica Applicata (IGA, Udine, Italy) (pools 1–5), the Center for Genome Research and Biocomputing (CGRB, Oregon State University, Corvallis, OR, USA) (pools 6–11), and IRTA (Centre de Recerca en Agrigenòmica CSIC-IRTA-UAB, Spain) (pool 12).

Pool	Accession	Adaptors	Read length (bp)	Read count (million)	Coverage of peach genome
1	'Armking'	CACAGT	94	5.85	2.42
1	'Big Top'	CGAGAT	94	3.55	1.47
1	'Fidelia'	ATGGCT	94	6.47	2.68
1	'Flordena'	GCATAG	94	7.27	3.01
1	'Silver Rome'	CATTCG	94	8.35	3.45
1	'Weinberger'	ACACTG	94	9.72	4.02
2	'Babygold 8'	TTGCGA	93	5.40	2.23
2	'Elberta'	CAGTAC	93	5.43	2.39
2	'Manaja'	TGCAAC	93	8.52	3.49
2	'Maycrest'	ACTAGC	93	8.61	3.52
3	'Oro A'	GAGCAA	93	7.20	2.95
3	'Stark Red Gold'	GCTACA	93	6.37	2.61
3	'Cica'	CATTCG	93	9.23	3.78
3	'Imera'	GCATAG	93	5.92	2.42
3	'Percoca di Romagna 7'	ATGGCT	93	4.27	1.75
3	'Pillar'	ACACTG	93	1.40	0.57
3	'S 2678'	CGAGAT	93	10.15	4.15
3	'Stark Saturn'	CACAGT	93	7.45	3.05
4	'Kamarat'	ACTAGC	93	9.63	3.94
4	'Leonforte 1'	GAGCAA	93	2.32	0.95
4	'Sahua Hong Pantao'	GCTACA	93	19.29	7.88
4	'Shen Zhou Mitao'	TTGCGA	93	12.54	5.13
4	'Tabacchiera'	TGCAAC	93	0.56	0.23
4	'Tudla'	CAGTAC	93	7.43	3.04
5	'GF677'[a]	CGAGAT	93	9.22	3.77
5	'Kurakata Wase'	ATGGCT	93	6.75	2.76
5	'Quetta'	CACAGT	93	12.76	5.22
5	'SM99'	ACACTG	93	4.90	2.01
6	'Admiral Dewey'	GGGT	80	2.42	0.85
6	'Babcock'	CCAT	80	3.19	1.12
6	'Elberta'	AGCT	80	0.64	0.23
6	'Slappey'	TCCT	80	2.02	0.71
7	'Bolinha'	AGCT	80	3.55	1.25
7	'Carmen'	GGGT	80	1.66	0.58
7	'Chinese Cling'	TCCT	80	2.50	0.88
7	'Mayflower'	CCAT	80	1.35	0.47
8	'Diamante'	AGCT	80	2.11	0.74
8	'J.H. Hale'	TCCT	80	3.18	1.12
8	'Rio Oso Gem'	CCAT	80	2.57	0.91
8	'Yellow St. John'	GGGT	80	1.35	0.48
9	'Dixon'	GGGT	80	1.25	0.44
9	'Early Crawford'	TCCT	80	3.89	1.37
9	'Florida Prince'	CCAT	80	1.85	0.65
9	'Nonpareil'[b]	AGCT	80	2.52	0.89
10	'Dr. Davis'	GGGT	80	2.31	0.81
10	'Nemaguard'	AGCT	80	2.38	0.84
10	'O'Henry'	TCCT	80	4.28	1.51
10	'Okinawa'	CCAT	80	2.15	0.76
11	'Georgia Belle'	AACT	80	14.42	5.08
11	'Lovell'	CTGT	80	6.55	2.30
11	'Lovell'	CCTT	80	0.03	0.01
11	'Oldmixon Free'	TGGT	80	3.26	1.15
12	'Big Top'	ACACGTAGTAT	330	0.28	0.28
12	'Brasco'	ACACGACGACT	355	0.16	0.20
12	'Catherina'	ACACTACTCGT	288	0.17	0.22
12	'Elegant Lady'	ACGACACGTAT	243	0.19	0.20
12	'Nectaross'	ACGAGTAGACT	275	0.19	0.23
12	'O'Henry'	ACGCGTCTAGT	289	0.15	0.18
12	'Sweet Cap'	ACGTACACACT	251	0.16	0.18
12	'Venus'	ACGTACTGTGT	278	0.15	0.19

[a]Peach x almond hybrid.
[b]Almond accession.
Pools 1–11 were sequenced with the Illumina Genome Analyzer while pool 12 was sequenced with the Roche 454 platform. Adaptors were used for retrieving accession-specific sequences from pools.
doi:10.1371/journal.pone.0035668.t001

doi:10.1371/journal.pone.0035668.t001

Aliquots of pooled libraries (5 pmol) were processed with the Illumina Cluster Generation Station, following the manufacturer's recommendations. Pools were sequenced in one lane of Illumina GA II with 94 cycles per read at the Istituto di Genomica Applicata (IGA, Udine, Italy) for pools 1–5 and with 80 cycles per read at the Center for Genome Research and Biocomputing (CGRB, Oregon State University, Corvallis, OR, USA) for pools 6–11, as specified in Table 1. Libraries for accessions in pool 12 were constructed for 454 GS FLX sequencing with MID-labeled libraries. Nuclear DNA from the eight accessions of pool 12 was digested with *Alu*1 and size-selected for 400–500 bp fragments. At IGA the CASAVA 1.7.0 version of the Illumina pipeline was used and at the OSU CGRB the CASAVA 1.6.0 version of the Illumina pipeline was used. Raw sequences were retrieved and kept separate for each accession and then aligned to the Peach v1.0 reference genome [20] using CLC Genomics Workbench (CLC Bio, Aarhus, Denmark) at IGA and *Soap2*[22] at the CGRB. In this paper, "chromosome" refers to one of the eight pseudomolecules (scaffolds) of the Peach v1.0 reference genome.

SNP detection

SNP detection and filtering followed a multi-step procedure (Figure 1). SNPs from sequences generated at IGA (pools 1–5) were detected using CLC Genomics Workbench (CLC Bio, Aarhus, Denmark), using default parameters for filtering except: (1) minimum Illumina quality score (Qscore) was 25; (2) minimum coverage of 30 reads and maximum coverage was 2.5× average coverage, corresponding to 161 reads; and (3) a minimum of five reads supported the presence of the minor allele in the accessions. SNPs in repetitive regions were also removed with internal scripts. SNPs from the 23 CGRB-sequenced accessions (pools 6–11) and from the eight accessions in pool 12 were detected using *SoapSNP*(http://soap.genomics.org.cn/soapsnp.html) essentially as recommended by Li [22]. In filtering for pools 6–12, SNPs were kept if: (1) the Illumina Qscore was more than 30 (except for pool 12 for which an Illumina Qscore could not be obtained); (2) the maximum number of reads for either allele across all accessions was less than the average read depth of all SNPs plus three standard deviations, 380 in this case; (3) a minimum of five reads supported the presence of the minor allele in the accessions, providing a minimum coverage of 10 reads for the SNP; and (4) the average copy number of the SNP flanking region was less than two, corresponding to non-repetitive regions of the genome. These detection and filtering efforts yielded "Stage 1 SNPs" (Figure 1).

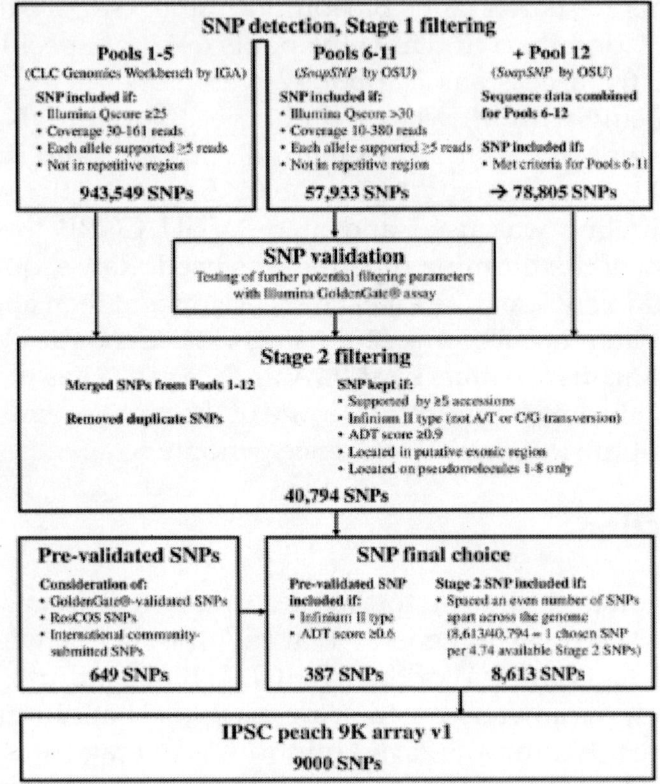

Figure 1. Workflow for SNP detection, validation, filtering, and final choice employed for development of the International Peach SNP Consortium (IPSC) peach 9 K SNP array v1.

doi:10.1371/journal.pone.0035668.g001

We compared the SNP calls in three of four pairs of accessions independently sequenced in different labs (Table 1). SNP calls were not compared for 'Lovell' because of low amount of sequence generated by one sample (0.01× coverage). For the remaining three cultivars, 'Elberta' (2.3× Illumina by IGA vs. 0.23× Illumina by

CGRB), 'Big Top' (1.47× Illumina by IGA vs. 0.29× 454 by IRTA), and 'O'Henry' (1.51× Illumina by CGRB vs. 0.18× 454 by IRTA) (Table 1) the SNP calls were compared when they belonged to the 320,747 SNPs supported by >4 accessions in Stage 2 filtering (Figure 1).

SNP validation with GoldenGate® assay

A set of 96 SNPs was chosen from the Stage 1 SNPs from pools 6–11 to validate the efficiency of SNP detection and adjust subsequent filtering parameters (Figure S1). The initial selection comprised 74 Stage 1 SNPs evenly spread over the eight pseudomolecules representing the haploid chromosomes and linkage groups (LGs) of the Peach v1.0 'dhLovell' genome assembly[20]. One SNP was chosen to be located within 200 kb of each end of each LG. SNPs chosen between these ends were then evenly spaced along each LG according to their total genetic distance [23], corresponding to one SNP every 2–5 Mb. The spacing across all LGs averaged 3.28 Mb (standard deviation of 0.67 Mb), with a minimum average of 2.51 Mb (±0.01 Mb) for LG7 and a maximum average of 4.44 Mb (±0.01 Mb) for LG2. Approximately 40% of the 74 chosen SNPs were in exons (CDS) of annotated genes, 20% in introns, 20% in 5′ or 3′ untranslated regions (UTR, outside genes but within 2 kb of start or stop codons), and the final 20% in intergenic regions. Fourteen more SNPs spanned a 774 kb region near the end of LG4, at a major trait locus associated with fruit texture (*Freestone-Melting flesh* [*F-M*] locus) [24], at an average spacing of 60 kb (ranging from 8 kb to 113 kb). The final eight validation SNPs targeted candidate genes for the *Y* locus on LG1 (four SNPs at 123, 270, and 341 kb intervals), the *Cs* locus on LG3 (two SNPs 38 kb apart), and the *G* locus on LG5 (two SNPs 272 kb apart) according to the genomic positioning reported by Dirlewanger [23]. While trait locus-targeted SNPs were chosen for variation in genic regions when possible, preference was given to achieving uniform target spacing in designated windows. Approximately 20% of the 96 SNPs chosen were planned to be accession-specific, i.e., their minor allele would be detected in only one re-sequenced accession of the detection panel (for which data available at the time included only accessions from pools 6–11). Sixteen of the evenly spaced SNPs and four of the trait locus-targeted SNPs met this criterion. Accession-specific SNPs were from nine peach accessions, with the almond

'Nonpareil' and the peach reference cultivar 'dhLovell' having five and six such SNPs, respectively. The 96 SNPs also deliberately included a wide range of MAFs.

To test the variables of genomic location, genic location, and MAF for their effects on genotyping efficiency, a validation panel of 160 *Prunus* accessions (54 peach cultivars, three interspecific hybrid cultivars, three almond cultivars, 59 breeding selections of peach and related species, and 41 seedlings of breeding populations as listed in Table S1) was screened with the 96-SNP subset, using the GoldenGate® assay. Individuals in the validation panel were founders, intermediate ancestors, important breeding parents, and seedlings of modern peach cultivars, forming a complex pedigree structure linking much of the world's cultivated peach crop and including the 23 accessions in pools 6–11 of the detection panel. The validation panel also included the T×E bin-mapping set of six 'Texas' almond×'Earlygold' peach seedlings and 'Earlygold' [25] to enable validation of SNP genomic locations. For a snapshot of SNP polymorphisms in breeding populations, five progenies of six F_1 seedlings were included, with one progeny set each from U.S. peach breeding programs in Arkansas, South Carolina, and Texas, and two from California. Genomic DNA was isolated from each accession using the E-Z 96 Tissue DNA Kit (Omega Bio-Tek, Inc., Norcross, GA, USA). DNA was quantitated with the Quant-iT™ PicoGreen® Assay (Life Technologies, Grand Island, NY, USA), using the Victor multiplate reader (Perkin Elmer Inc., San Jose, CA, USA). Concentrations were adjusted to a minimum of 50 ng/µl, in 5 µl aliquots and were submitted to the Research Technology Support Facility at Michigan State University (East Lansing, MI, USA), where the GoldenGate® assay was performed following the manufacturer's protocol (Illumina Inc.). After amplification, PCR products were hybridized to VeraCode microbeads via the address sequence, for detection on a VeraCode BeadXpress Reader. SNP genotypes were scored with the Genotyping Module of GenomeStudio Data Analysis software (Illumina Inc.).

SNP final choice

Stage 1 SNPs from pools 1–5 and pools 6–12 were independently converted to Illumina Assay Design Tool (ADT) format with custom scripts and scored by Illumina. "Infinium® I"-type SNPs (A/T or C/G transversions) were removed, as well as SNPs with any failure code or ADT score <0.9. Then the two datasets were merged, and duplicates were removed with custom scripts. The remaining SNPs were filtered by removing those: (1) where the last 25 bp of the 50 bp probe were duplicated; (2) supported by less than five accessions; (3) not in predicted coding regions; and (4) not located on one of the first eight pseudomolecules representing the eight peach chromosomes. This process yielded "Stage 2 SNPs" (Figure 1).

Pre-validated SNPs were obtained from several sources: (1) polymorphic SNPs from the GoldenGate® validation activity; (2) peach RosCOS SNPs [26]; and (3) SNPs requested for inclusion by the international peach genomics community. Pre-validated SNPs were filtered to remove Infinium® I SNP types and those with an ADT score <0.6.

For the final choice of 9,000 SNPs for the IPSC peach 9 K SNP array v1, filtered pre-validated SNPs were automatically included. For the remainder, SNPs were chosen to provide an even spacing across the genome, corresponding to one SNP selected for every 4.74 Stage 2 SNPs (40,794/8,613).

SNP array evaluation

The IPSC peach 9 K SNP Infinium® II array v1 was evaluated using 709 accessions divided in two independent evaluation panels, one panel from the European Union (EU) and the other from the USA (US). The EU panel comprised 232 accessions, of which 229 were peach cultivars and three were wild related *Prunus* species or their hybrids with peach. The US panel comprised 479 accessions that included pedigree-linked cultivars, breeding lines, and seedlings (Table S2). Overall, selected material comprised cultivars (45%), advanced selections (4%) and seedlings (51%). Accessions with pure peach and almond ancestry accounted for 82% and 2%, respectively, while 16% of genotyped material had interspecific backgrounds with almond (7%), and peach and almond wild relatives, 5% and 4%, respectively, in their pedigrees. Some US panel accessions were related *Prunus* species or were known interspecific hybrids: 5% had peach-related (*P. davidiana* and *P. mira*) ancestry, 10% had almond (*P. dulcis*), and 3% had almond-related (*P. argentea* and *P.*

scoparia) ancestry. Genomic DNA extraction and quantitation were conducted as described above for the SNP validation panel for the U.S. accessions. For the EU panel, genomic DNA was extracted using the DNeasy Plant Mini Kit (Qiagen) and quantitated using a Fluoroskan Ascent (Thermo Scientific, Finland) microplate reader. The IPSC array, employing exclusively Illumina Infinium® II design probes and dual color channel assays (Infinium HD Assay Ultra, Illumina), was used for genotyping, following the manufacturer's recommendations. SNP genotypes were scored with the Genotyping Module of the GenomeStudio Data Analysis software (Illumina, Inc.). A GenTrain score of >0.4 and a GenCall 10% of >0.2 were applied to remove most SNPs that did not cluster (homozygous), or had ambiguous clustering. SNPs that did not cluster for more than 50% of samples were also eliminated from further consideration. The threshold of allowed No Calls (failed genotyping) was 'relaxed' in anticipation of the presence of null alleles for some SNPs contributed by non-peach species.

RESULTS

SNP detection

A total of 25.4 Gb of DNA sequence (111.7× coverage of the peach genome) was obtained from 279.7 million reads (Illumina and 454) generated for the 56 peach accessions multiplexed among 12 pools (Table 1). Excluding one sequencing run for 'Lovell' that generated an unusually low number of reads, the total number of reads sequenced using the Illumina platform averaged 2.16× coverage per accession and ranged from ~0.56 to 19.2 million reads in 'Tabacchiera' and 'Sahua Hong Pantao', respectively. The total number of reads per accession sequenced using the 454 platform averaged 0.22× coverage and ranged from 0.145 to 0.198 million reads in 'Sweet Cap' and 'Big Top', respectively. The number of SNPs identified after Stage 1 detection was 943,549 for pools 1–5 and 57,933 for pools 6–11 (Figure 1). When the same filtering parameters as those used for pool 6–11 were reapplied to pools 6–12, the number of SNPs increased to 78,805.

When the SNP calls were compared among the three pairs of

independently sequenced accessions, the majority of the positions were not covered by at least one of the two datasets and corresponded to 297,078, 301,617 and 309,027 positions in 'Elberta', 'Big Top' and 'O'Henry', respectively. Thus, the SNP positions that could be compared among these accessions ranged from 3.75% and 4% of the SNP calls in 'O'Henry' and 'Big Top' (sequenced by Illumina and 454) to 7.48% in Illumina-resequenced 'Elberta'. In these positions, most of the SNPs had the same genotype (51.2%, 72.4% and 79.9% of reads in 'Elberta', 'Big Top' and 'O'Henry', respectively) while a negligible number of positions did not share any allele (0.05% of positions in 'Big Top' to 0.6% in 'Elberta' and 'O'Henry').

SNP validation

Several SNP characteristics were associated with performance differences in the GoldenGate® assay (Table 2). Exonic and intronic SNPs were the most successful, with approximately 75% of polymorphisms verified; in contrast, intergenic SNPs were the worst performers, with only a third being polymorphic and a third failing (Table 2, Figure S2). Although MAF observed from sequencing of the detection panel (n = 23) and MAF observed after GoldenGate® genotyping of the validation panel (n = 119) were not well correlated (R = 0.12), the higher the detection panel MAF of a SNP, especially >30%, the more likely its validation panel MAF was >10%. Yet SNPs with a detection panel MAF of <20% were more likely than higher MAF SNPs to convert to a validation panel MAF of <10% without a higher-than-average rate of failure or monomorphism (Table 2). Of the 74 SNPs evenly distributed over the genome, those on LG6, LG1, and LG3 had the highest conversion to polymorphism yet did not have a higher-than-average proportion of exonic and intronic SNPs (Table 2, Figure S2). LG1 and LG6 also had the highest SNP representation because they were the longest in genetic length. LG8 gave a high proportion of polymorphic SNPs with low MAF (<10% in the validation panel). LG2 had a very high failure rate yet only an average proportion of exonic and intronic SNPs (Table 2, Figure S2). Accession-specific SNPs had a close to average performance for rate of failure, monomorphism, and polymorphism, but provided a higher-than-average proportion of low MAF SNPs observed in the validation panel. No SNPs designated as accession-specific in the detection panel remained accession-specific after genotyping the validation panel, although those developed from the almond accession 'Nonpareil' were obviously indicative of almond introgression in the validation panel. SNPs targeted to the *F-M* locus also provided a high proportion of low-MAF SNPs, but had a high failure and monomorphism rate as did the SNPs targeted to other trait loci (Table 2). All trait locus-targeted SNPs were associated with a very high proportion of intergenic SNPs (Figure S2). Overall, SNP failure was similar across all parameters except for a high level for SNPs that were on LG2, targeted to trait loci, or intergenic. Monomorphism was highest for SNPs with a detection panel MAF of 21–30%, trait locus-targeted, or those that were intergenic (Table 2).

Table 2. Validation outcomes for 96 SNPs with the GoldenGate® assay.

SNP parameter	Total	Proportion of SNPs					
		Failed	Monomorphic	Polymorphic	MAF (validation panel)		
					<5%	5–10%	>10%
Evenly spaced	74	0.23	0.14	0.63	0.01	0.04	0.59
F-M locus	14	0.29	0.29	0.36	**0.07**	**0.14**	0.14
Other trait loci	8	0.38	0.38	0.25	0.00	0.00	0.25
Accession-specific	19	0.26	0.11	0.63	**0.11**	**0.11**	0.42
Genomic location:							
LG1	13	0.08	0.15	**0.77**	0.00	0.00	**0.77**
LG2	7	0.71	0.00	0.29	0.00	0.00	0.29
LG3	7	0.14	0.14	**0.71**	0.00	0.00	**0.71**
LG4	9	0.33	0.11	0.56	0.00	0.11	0.44
LG5	7	0.14	0.29	0.57	0.00	0.00	0.57
LG6	12	0.00	0.17	**0.83**	0.00	0.00	**0.83**
LG7	10	0.30	0.10	0.60	0.00	0.00	0.60
LG8	9	0.22	0.11	0.67	**0.11**	**0.22**	0.33
Genic location:							
Exonic	30	0.10	0.13	**0.77**	0.00	0.03	**0.73**
Intronic	19	0.21	0.05	**0.74**	0.00	0.05	**0.68**
UTR	16	0.25	0.25	0.50	0.00	0.00	0.50
Intergenic	31	0.48	0.12	0.42	0.06	0.10	0.16
ADT score:							
<0.2	2	1.00	0.00	0.00	0.00	0.00	0.00
0.2–0.4	4	1.00	0.00	0.00	0.00	0.00	0.00
0.4–0.5	2	0.00	0.50	0.50	0.00	0.00	0.50
0.5–0.6	1	0.00	0.00	1.00	0.00	0.00	1.00
0.6–0.7	8	0.38	0.13	0.50	0.00	**0.13**	**0.38**
0.7–0.8	10	0.20	0.30	0.50	**0.10**	0.00	0.40
0.8–0.9	18	0.17	0.22	0.61	0.06	0.00	**0.56**
>0.90	54	0.22	0.17	0.61	0.00	0.07	**0.54**
MAF (detection panel):							
1–10%	8	0.25	0.25	0.50	0.00	**0.13**	0.38
11–20%	20	0.25	0.15	0.60	**0.10**	**0.15**	0.35
21–30%	9	0.00	0.56	0.44	0.00	0.00	0.44
31–40%	25	0.24	0.20	0.56	0.00	0.04	**0.52**
41–50%	34	0.29	0.09	0.62	0.00	0.00	**0.62**
Total	96	0.24	0.19	0.57	0.02	0.05	0.50

Observed minor allele frequency (MAF) of polymorphic SNPs among 23 accessions of the detection panel and 119 non-seedling accessions of the validation panel are indicated. Values in bold represent considerably better than average SNP performance (e.g., high polymorphism), while values in italics are worse than average (high failure and monomorphism).

doi:10.1371/journal.pone.0035668.t002

doi:10.1371/journal.pone.0035668.t002

Of the 32 validation SNPs that were polymorphic in the interspecific T×E bin-mapping set, 31 had joint genotypes corresponding to their expected genomic region. The presence of null alleles was evident for 14 of these polymorphic SNPs. One intergenic SNP targeted to the *F-M*locus on LG4 (bin 4:63; Scaffold_4:22274908) resulted in polymorphism (MAF 23%) with a joint genotype that unexpectedly placed it on LG2 (bin 2:13).

Of the 55 SNPs polymorphic in the entire validation panel, 52 (95%) were polymorphic in at least one of the five six-seedling breeding progenies. One SNP was polymorphic in all five progenies, 10 in four progenies, 17 in three progenies, 10 in two, and 14 in just one progeny. Proportions of the 55 polymorphic SNPs that were polymorphic within individual progenies were 58%, 40%, 53%, 44%, and 42% from Arkansas (University of Arkansas Fruit Research Station, Clarksville), California (National Clonal Germplasm Repository for Fruit and Nut Crops at Davis, CA), California (UC Davis Wolfskill Orchard at Winters, CA), South Carolina (Clemson University Musser Fruit Research Station, Clemson, SC), and Texas (Texas A&M University-College Station Research farm), respectively. These proportions were lower bounds, given that the presence of segregating null alleles was not considered for these progenies.

SNP final choice

Stage 2 filtering reduced the one million available SNPs by 96%. Restricting the SNPs to Infinium® II types eliminated 18% of Stage 1 SNPs from consideration, the thresholds of ADT≥0.9 and supporting evidence from at least five accessions eliminated 50% of Stage 1 SNPs, and restriction to genic regions cut an additional 23%. Considering only exonic SNPs more than halved the remaining 86,360 to 41,800. Finally, 1006 SNPs located on peach genome scaffolds as yet unassigned to one of the eight chromosomes of this crop were discarded, resulting in 40,794 Stage 2 SNPs (Figure 1, Table 3). The distribution of Stage 2 SNPs varied among and within the eight peach chromosomes (Figure 2, Table 3).

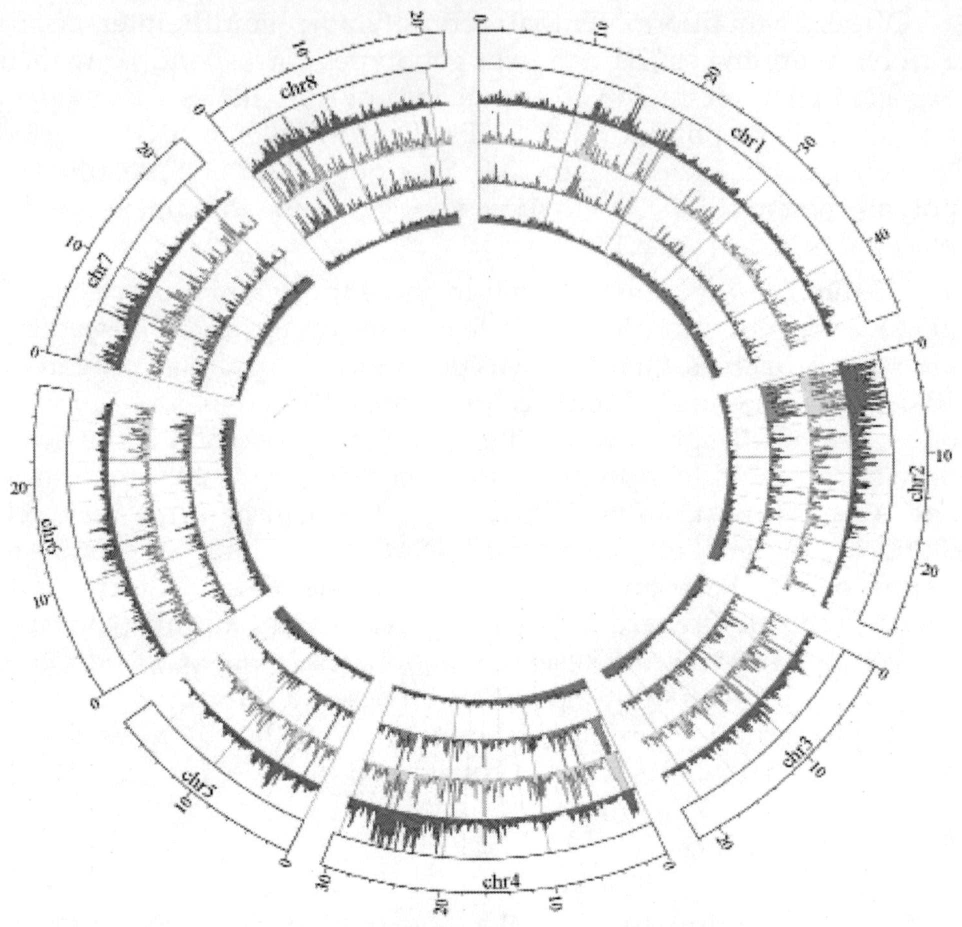

Figure 2. Distribution of SNPs along the Peach v1.0 pseudomolecules.

All tracks are plotted in 100 kb windows; inner blue track represents the frequency of coding DNA sequence CDS; y axis ranges from 0 to 100%. Red, yellow and green tracks represent, respectively, absolute number of SNPs discovered within pool 1–5, 40,789 Stage 2 SNPs in exons, and 9,000 SNPs chosen for the array; values in the y axes are capped at 2000, 100, and 30, respectively.

doi:10.1371/journal.pone.0035668.g002

Table 3. Chromosome distribution and performance of SNPs on the IPSC peach 9 K SNP array v1.

Chromosome	Stage 2 SNPs	SNPs on peach array	SNPs polymorphic on the peach array	Rate of Polymorphism (%)	Distance between SNPs		
					Average gap (kb)	Largest gap (kb)	No. of gaps >150 kb
1	[illegible]	[illegible]	[illegible]	[illegible]	[illegible]	[illegible]	[illegible]
2	[illegible]	[illegible]	[illegible]	[illegible]	[illegible]	[illegible]	[illegible]
3	[illegible]	[illegible]	[illegible]	[illegible]	[illegible]	[illegible]	[illegible]
4	[illegible]	[illegible]	[illegible]	[illegible]	[illegible]	[illegible]	[illegible]
5	[illegible]	[illegible]	[illegible]	[illegible]	[illegible]	[illegible]	[illegible]
6	[illegible]	[illegible]	[illegible]	[illegible]	[illegible]	[illegible]	[illegible]
7	[illegible]	[illegible]	[illegible]	[illegible]	[illegible]	[illegible]	[illegible]
8	[illegible]	[illegible]	[illegible]	[illegible]	[illegible]	[illegible]	[illegible]
Total on chromosomes	[illegible]	[illegible]	[illegible]	[illegible]	[illegible]	[illegible]	[illegible]
Unanchored scaffolds	[illegible]	[illegible]	[illegible]	[illegible]			
Total	[illegible]	[illegible]	[illegible]	[illegible]			

[illegible]

doi:10.1371/journal.pone.0035668.t003

A total of 649 pre-validated SNPs considered for inclusion in the final array were reduced to 387 by the filtering parameters used for these SNPs. Of the 55 polymorphic SNPs from the GoldenGate® validation assay, 45 were included. Of the 453 peach RosCOS SNPs considered, 225 were included in the final array. SNPs requested for inclusion from the international community were 108 from the DRUPOMICS project (Verde I., unpublished data) and 33 from the University of Chile (Silva H., unpublished data), of which 87 and 30 were included, respectively. Nineteen of the 387 pre-validated SNPs were located on unanchored scaffolds of the Peach v1.0 assembly, four of which were RosCOS SNPs. Inclusion of the 387 pre-validated SNPs left 8,613 positions available in the final array for SNPs distributed over the eight peach chromosomes.

The IPSC peach 9 K SNP array v1 achieved an average spacing of 26.7 kb between SNPs (Table 3). Distribution of SNPs along the Peach v1.0 pseudomolecules varied according to the number of Stage 2 SNPs observed throughout the genome (Figure 2). The largest average gap between successive SNPs (42.0 kb) occurred on chromosome 1 and the smallest on chromosome 4 (18.6 kb). A total of 252 gaps were larger than 150 kb and the largest single gap was 915.8 kb on chromosome 5 (Table 3), but the vast majority of gaps were less than 5 kb (Figure 3). For most of the regions with large gaps there were simply no SNPs available to reduce the gap size considerably. However some of the gaps were caused by the loss of 856 SNPs that occurred during the manufacturing of the array.

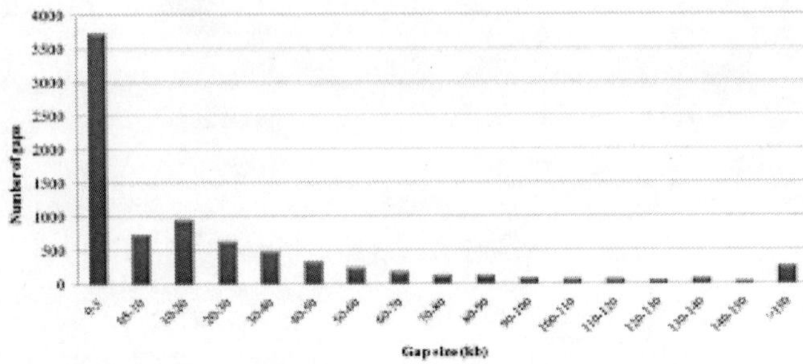

Figure 3. Frequency distribution of size of gaps between SNPs included on the IPSC peach 9 K SNP array v1. Gap sizes were based on SNP physical locations in the Peach v1.0 assembly.

doi:10.1371/journal.pone.0035668.g003

SNP array evaluation

Of the 9,000 candidate SNPs, 8,144 remained on the array after Illumina technical dropout (loss during array manufacturing). Of these, 8,125 were located on the first eight pseudomolecules of the peach genome. The evaluation of IPSC peach 9 K SNP array v1, performed in Europe and U.S., revealed no significant difference in the number of polymorphic, monomorphic and failed SNPs between peach-only samples and samples with interspecific backgrounds (data not shown). The only difference was observed for SNPs located in RosCOS, with only 2% of them being polymorphic in peach-only samples versus 6% in all samples, including those with interspecific ancestry. Moreover, independent evaluation of the IPSC peach 9 K SNP array v1 using EU and US accession panels revealed almost identical results for the numbers of polymorphic, monomorphic and failed SNPs. Genome scans performed with the IPSC peach 9 K SNP array v1 on the EU and US evaluation panels resulted in a SNP polymorphism rate of 70% for each panel and a failure rate of only 5% for each (Table S3). Of the polymorphic SNPs, 92% were observed with a MAF>0.10 for each panel, and 74% and the 69% had a MAF>0.20 in the EU and US evaluation panels, respectively (Table S3). When the US evaluation panel was reduced to only cultivars and advanced selections, the proportion of MAF>0.20 (73%) was almost identical to that observed in the EU evaluation panel. Proportions of SNPs in each MAF range were similar between the two evaluation panels regardless of the type of the material analyzed (Figure 4). In addition to 5,967 SNPs (73%) polymorphic in both evaluation panels, 425 (5%) and 477 (6%) were polymorphic only in the EU and US panels, respectively (Figure 5).

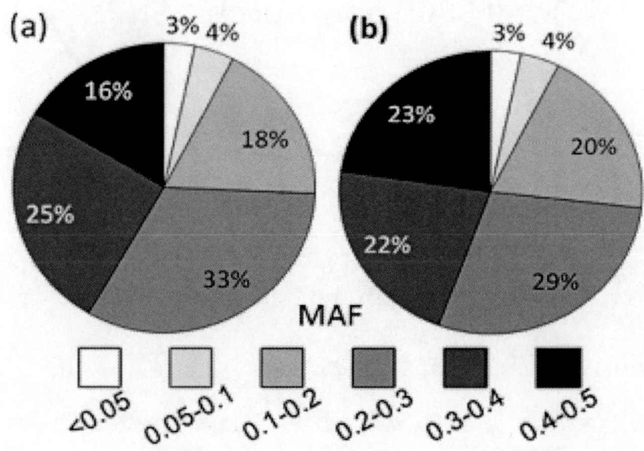

Figure 4. Distribution of minor allele frequencies (MAF) in two independent germplasm sets.

A. EU evaluation panel (n = 232); B. US evaluation panel (n = 115; cultivars and advanced selections only).

doi:10.1371/journal.pone.0035668.g004

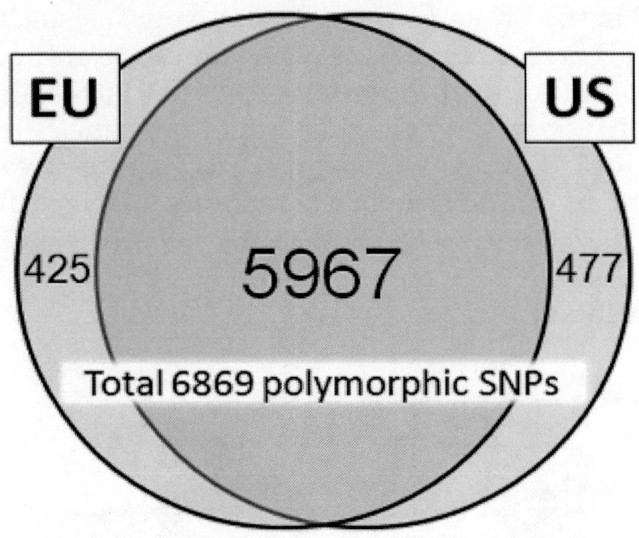

Figure 5. Polymorphic SNPs detected in EU (n = 232) and US (n = 477) evaluation panels from genome scans with the IPSC peach 9 K SNP array v1.

doi:10.1371/journal.pone.0035668.g005

The IPSC peach 9 K SNP array v1 achieved a total of 6,869 polymorphic SNPs across the 709 accessions scanned in the two panels combined (Table 3), representing 84.3% of SNPs present on the array. These polymorphic markers provided an average spacing of 31.5 kb across the peach genome, which was consistent among the chromosomes. Fourteen SNPs out of the nineteen localized on the unanchored scaffolds of Peach v1.0 assembly were polymorphic. They provide useful coverage for the unmapped fraction of the Peach v1.0 assembly by helping to reduce gaps. The largest gaps between polymorphic SNPs were on chromosome 1 (1,254 kb). Physical positions and MAFs of polymorphic SNPs were compared for the US peach and non-peach cultivars and selections (Figure 6). The coefficient of regression (r) between MAF in "peach cultivars and selections" and in "non-peach cultivars and selections" within the US panel was 0.437. Predicted SNPs in RosCOS loci were mostly monomorphic, with only 14 being polymorphic.

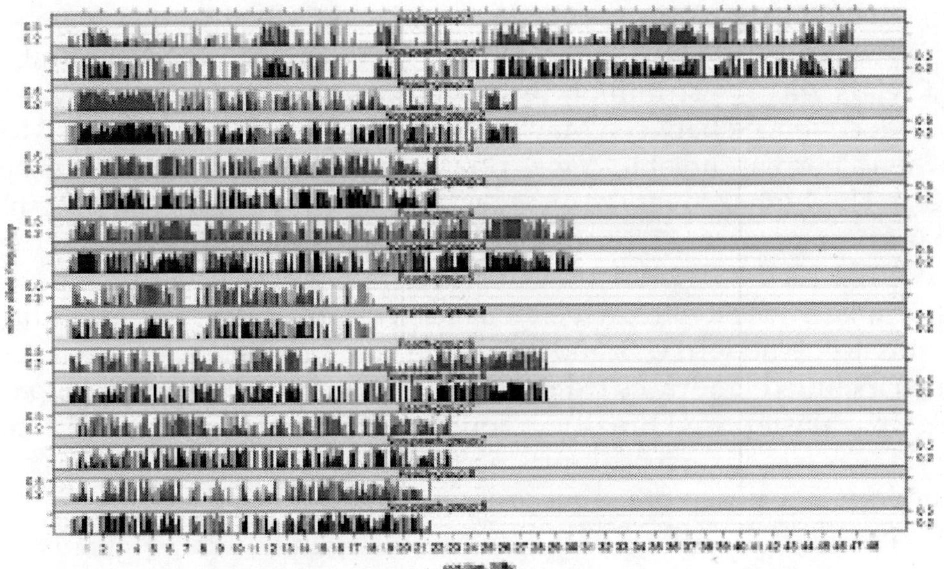

Figure 6. Distribution and physical spacing of polymorphic SNPs across the eight peach chromosomes and comparison of SNP minor allele frequencies between peach and non-peach samples, including almond, and peach and almond wild relatives, in US data set.

Data set comprises cultivars and advanced selections only. The coefficient of regression (r) between the MAF in peach and non peach set is 0.437.

doi:10.1371/journal.pone.0035668.g006

DISCUSSION

SNP detection

The large difference in the number of SNPs identified between pools 1–5 and pools 6–11 is not surprising given the much higher sequencing coverage of pools 1–5 (84.8×) than 6–11 (25.1×) (Table 1) and the greater genetic diversity expected in pools 1–5. Updates in software and hardware for the Illumina Genome Analyzer during the study were major reasons for the difference in coverage between the two main pool groups. Accessions in pools 6–11 were sequenced in February 2010 with GAIIx technology, when the expected number of reads was 17–21 million reads per lane. Accessions in pools 1–5 were sequenced between May and June 2010 with an Illumina GAIIx, when a normal run at IGA was producing about 60 million reads per lane. This higher coverage in pools 1–5 and the SNP validation results based on pools 6–11 prompted some adjustments to the filtering adopted for SNP prediction in pools 1–5. Modern western peach germplasm has a narrow genetic base [15]. In addition to including cultivars of modern U.S. and European breeding programs, pools 1–5 also included landraces from around the world (China, Italy, Japan, Spain, Pakistan, and Brazil). In contrast, the set of accessions of pools 6–11 was less diverse, comprising mostly founders of importance to modern U.S. peach breeding programs.

The low overall percentage of SNPs with matching genotypes in the three independently sequenced pairs of accessions is caused by low sequence coverage targeted here. The sequences generated in this project that were not included in the IPSC peach 9 K SNP array v1 are a valuable resource for peach and other *Prunus* species as they can be searched for additional sequence variation in regions underlying traits of economic importance.

SNP VALIDATION

SNP validation was a critical step for empirically determining appropriate parameters for prioritizing detected SNPs, improving the SNP detection for pools 1–5 that was conducted after SNP detection for pools 6–11, and to choose from among approximately one million SNPs just 9,000 used in the final array design. The validity of the GoldenGate® genotyping results for extrapolation of parameter thresholds to all detected SNPs was supported by the correct genomic location assignment of polymorphic SNPs by bin-mapping. Our observation that conversion of NGS-detected SNPs to true SNPs was more reliable for polymorphisms located in predicted exons and introns was consistent with observations in apple [6]. Better performance of genic SNPs may be due to a lower level of undetected sequence variation in SNP-flanking regions, as opposed to reduced sequence conservation in UTRs and intergenic space, as argued by Chagné et al. [6]. If so, deeper sequencing of detection panel accessions to identify a larger proportion of the polymorphisms would enable avoidance of SNPs with polymorphisms in flanking probe sequences and hence increase the probability of successful SNP marker development. The low depth of sequencing of individual accessions in the detection panel is the probable reason for the lack of significant correlation between detection panel MAF and validation panel MAF. However, detection panel MAF was still a useful parameter for developing rare (<10%) or common (>10%) SNPs. Rare SNPs would be useful for detecting unique haplotypes in a germplasm set, although they would be uninformative for most individuals. Common SNPs would be useful for detecting the majority of genetic variation in a germplasm set and, with a high degree of saturation of any given genomic region, may detect most or all unique haplotypes and thus overcome a lack of rare lineage-specific SNPs. The accession specificity of a SNP in the detection panel was found to be a redundant criterion for developing rare lineage-specific SNPs, and in any case, the relatively small number of accessions in the detection panel and low sequencing coverage of each negated true specificity.

The reason for differential performance of SNPs among some LGs is not clear. In the development of a genome-wide array, it is not advisable to avoid certain LGs or to saturate others just to increase the overall proportion of successful SNPs. These differences were not maintained in the final array (Table 3) and hence probably represent sampling error in the relatively small set of validation SNPs.

Although only about half of the successful SNPs segregated in any of the four individual F_1 populations included in the validation assay, together these four populations captured most (95%) of the polymorphisms. This observation highlights the importance of utilizing diverse target germplasm rather than a narrow subset of the target genepool, such as a single population, for SNP evaluation and assessment of array informativeness.

SNP FINAL CHOICE

The SNPs identified within this internationally coordinated effort were subjected to stringent filtering criteria to maximize the efficacy of the IPSC peach 9 K SNP array v1. Given that most candidate

SNPs were essentially anonymous, Infinium® I SNPs were avoided because these require two bead types and thus use two SNP slots of the assay, unlike the single bead type for Infinium® II SNPs. Infinium® II SNPs have also been reported to perform better (92% vs. 85%) [3]. Furthermore, Infinium® II SNPs were the most abundant class detected in our study:

82% of the SNPs detected were of this type. SNPs with an ADT score higher than 0.9 were used to increase the probability of SNP success as observed in the validation assay. This Illumina design score reflects the predicted ability of the SNP-flanking sequences to provide a successful assay [5]. The availability of a large number of SNPs allowed us to employ a stringent threshold that was higher than that used in the recently released Illumina Beadchips for apple, pig and chicken, which used ADT thresholds of 0.7, 0.8 and 0.6, respectively [6], [4],[5].

An important aspect we took into consideration in choosing the SNPs for the array was their genomic context. The large number of SNPs discovered in this study allowed us to select 9,000 exonic markers to be included in the array. The high level of transferability of transcriptomic markers across *Prunus* species [27] makes these exonic SNPs useful tools even for related non-peach species such as almond. We also expected exonic SNPs to be more commonly associated with causative mutations underlying phenotypic differences than intergenic SNPs, because of their potential to alter protein sequences.

Finally, another important criterion for a whole genome genotyping assay is to have a uniform distribution of the SNPs across the genome, as this greatly facilitates finding associations between markers and phenotypes. The SNPs on our array cover most of the peach genome with markers well distributed over all chromosomes. The average gap size across the genome achieved was 26.7 kb and increases to 31.5 kb when only polymorphic SNPs are considered. The average ratio of genetic to physical distance in peach is about 440 kb/cM, which was obtained by comparing the Peach v1.0 assembly with an updated version of the *Prunus*reference map [23] (IPGI unpublished results), and this gives an average of 13.3 polymorphic SNPs per cM for our array. Such resolution provides unprecedented power to dissect the peach genome for pinpointing QTLs and determining genetic relatedness. However, SNPs on the array developed here were not evenly spaced across the genome physically or genetically, and a few regions remained with gaps up to 915.8 kb (1,254 kb when considering only polymorphic markers). As the SNP spacing on the array was determined from the density of SNPs detected across the genome and was randomly but proportionally reduced by filtering and non-polymorphism, we expect that the largest gaps exist in regions of low polymorphism (low informativeness) in this crop. Even the largest gap corresponds to a genetic distance of about 2.8 cM, which is still powerful for QTL and relatedness studies. LD decay in peach was estimated at about 13–15 cM [14], so even the regions with large gaps have a density of markers at least 5-fold higher than estimated to be needed to perform optimal GWAS in peach.

SNP array evaluation

In summary, the total of 6,869 SNPs on the IPSC peach 9 K SNP array v1 verified as polymorphic through extensive empirical evaluation represent an excellent source of markers for studies in genetic relatedness, genetic mapping, and dissecting the genetic architecture of complex agricultural traits. The SNPs included in the array were successfully used for genotyping 709 accessions in two independent evaluation panels. The majority (86.9%) of these markers were polymorphic in both experiments (Figure 5) indicating that the array contains a very low number of false positive SNPs. The comparison (Figure 6) between peach and non-peach accessions showed a common pattern in MAF distribution and large common gaps. The correlation (r = 0.437) between MAF calculated within peach vs. non peach cultivars and selections shows that the array can be efficiently used in peach related species as well. The most notable difference in MAF for the polymorphic SNPs between the peach and non-peach selections was on chromosome 2 around 19 Mbp, where the non-peach group had higher MAF. The common gaps on chromosome 1 (~21 Mbp), chromosome 2 (~8.5 Mbp), chromosome 4 (~24.5 Mbp), chromosome 5 (~7 Mbp), and chromosome 8 (~10 Mbp), may represent putative centromeric regions (S. Scalabrin, personal communication).

CONCLUSION

The SNP array described here will foster genetic studies in the stone fruits and will help bridge the gap between genomics and breeding activities because breeding germplasm was the basis of detected SNPs and SNP choices of the final array. The IPSC peach 9 K SNP array v1 is commercially available and we expect that it will be used worldwide for genetic studies in peach and related stone fruit and nut species.

ACKNOWLEDGMENTS

We acknowledge the International Peach Genome Initiative for pre-publication access to the peach genome sequence. Peach breeders John Clark (University of Arkansas, Arkansas), Tom Gradziel (UC Davis, California), and Dave Byrne (Texas A&M University, Texas) helpfully provided leaf material for validation panel accessions. Daniele Bassi (DiProVe University of Milan, Italy) and Raffaele Testolin (University of Udine, Italy) kindly helped in the choice of the Pool 1–5 set and Daniele Bassi kindly provided leaf material for the resequencing of two peach accessions. We thank Claudio Cantini at Ivalsa-CNR Follonica, Italy and Szani Zsolt at Central Agricultural Office Budapest, Hungary for providing leaf material used in array evaluation. We are indebted with Maria Teresa Dettori at CRA – Fruit Tree Research Center for her critical advice in data analysis. We are grateful to Chris Sullivan, Mark Dasenko, and Matthew Peterson at the OSU Center for Genome Research and Biocomputing (CGRB) for their help in sequencing and data processing. We thank all international researchers who participated in the Rosaceae SNP Summit, Stellenbosch, South Africa, November 11–12th 2010, for their contribution to international needs assessment and design of SNP filtering criteria.

AUTHOR CONTRIBUTIONS

Conceived and designed the experiments: IV NB PA AI MM CP. Performed the experiments: NB BG CTL FC EV VA ALB MT. Analyzed the data: SS KG D. Micheletti D. Main CTL UR FC IV TCM

DWB MJA AI LW. Contributed reagents/materials/analysis tools: IV NB CTL PA AI BS MM CP TCM DWB. Wrote the paper: IV NB SS KG D. Micheletti CP TCM UR.

REFERENCES

1. Gunderson KL, Steemers FJ, Lee G, Mendoza LG, Chee MS (2005) A genome-wide scalable SNP genotyping assay using microarray technology. Nature Genetics 37: 549–554.
2. Gunderson KL (2009) Whole-genome genotyping on bead arrays. Meth Mol Biol 529: 197–213.
3. Matukumalli LK, Lawley CT, Schnabel RD, Taylor JF, Allan MF, et al. (2009) Development and Characterization of a High Density SNP Genotyping Assay for Cattle. PLoS ONE 4: e5350.
4. Ramos AM, Crooijmans RPMA, Affara NA, Amaral AJ, Archibald AL, et al. (2009) Design of a High Density SNP Genotyping Assay in the Pig Using SNPs Identified and Characterized by Next Generation Sequencing Technology. PLoS ONE 4(8): e6524.
5. Groenen MA, Megens HJ, Zare Y, Warren WC, Hillier LW, et al. (2011) The development and characterization of a 60 K SNP chip for chicken. BMC Genomics 12: 274.
6. Chagné D, Crowhurst RN, Troggio M, Davey MW, Gilmore B, et al. (2012) Genome-wide SNP detection, validation, and development of an 8 K SNP array for apple. PloS ONE 7(2): e31745. doi:10.1371/journal.pone.0031745.
7. Dirlewanger E, Moing A, Rothan C, Svanella L, Pronier V, et al. (1999) Mapping QTL controlling fruit quality in peach (*Prunus persica* (L.) Batsch). Theor Appl Genet 98: 18–31.
8. Etienne C, Rothan C, Moing A, Plomion C, Bodenes C, et al. (2002) Candidate genes and QTL for sugar and organic acid content in peach (*Prunus persica* (L.) Batsch). Theor Appl Genet 105: 145–159.
9. Verde I, Quarta R, Cerdrola C, Dettori MT (2002) QTL analysis of agronomic traits in a BC1 peach population. Acta Hortic 592: 291–297.
10. Quilot B, Wu BH, Kervella J, Génard M, Foulongne M, et al. (2004) QTL analysis of quality traits in an advanced backcross between *Prunus persica* cultivars and the wild relative species *P. davidiana*. Theor Appl Genet 109: 884–897.
11. Decroocq V, Foulongne M, Lambert P, Le Gall P, Mantin C, et al. (2005) Analogues of virus resistance genes map to QTLs for resistance to sharka disease in *Prunus davidiana*. Mol Genet Genomics 272: 680–689.

12. Cantín CM, Crisosto CH, Ogundiwin EA, Gradziel T, Torrents J, et al. (2010) Chilling injury susceptibility in an intra-specific peach [*Prunus persica* (L.) Batsch] progeny. Postharvest Biol Tech 58: 79–87.
13. Fan S, Bielenberg DG, Zhebentyayeva TN, Reighard GL, Okie WR, et al. (2010) Mapping quantitative trait loci associated with chilling requirement, heat requirement and bloom date in peach (*Prunus persica*). New Phytol 185: 917–930.
14. Aranzana MJ, Abbassi el-K, Howad W, Arús P (2010) Genetic variation, population structure and linkage disequilibrium in peach commercial varieties. BMC Genet 11: 69.
15. Scorza R, Mehlenbacher SA, Lighner GW (1985) Inbreeding and coancestry of freestone peach cultivars of the eastern United States and implications for peach germplasm improvement. J Am Soc Hortic Sci 110: 547–552.
16. Yoon JH, Liu DC, Song WS, Liu WS, Zhang AM, et al. (2006) Genetic diversity and ecogeographical phylogenetic relationships among peach and nectarine cultivars based on simple sequence repeat (SSR) markers. J Amer Soc Hort Sci 131: 513–521.
17. Barnaud A, Lacombe T, Doligez A (2006) Linkage disequilibrium in cultivated grapevine*Vitis vinifera* L. Theor Appl Genet 112: 708–716.
18. Lijavetzky D, Cabezas JA, Ibáñez A, Rodríguez V, Martínez-Zapater JM (2004) High throughput SNP discovery and genotyping in grapevine (*Vitis vinifera* L.) by combining a re-sequencing approach and SNPlex technology. BMC Genomics 8: 424.
19. Remington DL, Thornsberry JM, Matsuoka Y, Wilson LM, Whitt SR, et al. (2001) Structure of linkage disequilibrium and phenotypic associations in the maize genome in the maize genome. Proc Natl Acad Sci USA 98: 11479–11484.
20. International Peach Genome Initiative. http://www.rosaceae.org/peach/genome.
21. Marchese A, Tobutt KR, Caruso T (2005) Molecular characterization of Sicilian *Prunus persica* cultivars using microsatellites. J Hort Sci & Bio 80(1): 121–127.
22. Li R, Li Y, Kristiansen K, Wang J (2008) SOAP: short oligonucleotide alignment program. Bioinformatics 24(5): 713–714. doi:10.1093/bioinformatics/btn025.
23. Dirlewanger E, Graziano E, Joobeur T, Garriga-Calderé F, Cosson P, et al. (2004) Comparative mapping and marker-assisted selection in Rosaceae fruit crops. Proc Natl Acad Sci USA 101: 9891–9896.

24. Ogundiwin EA, Peace CP, Gradziel TM, Parfitt DE, Bliss FA, et al. (2010) A fruit quality gene map of *Prunus*. BMC Genomics 10: 587.

25. Howad W, Yamamoto T, Dirlewanger E, Testolin R, Cosson P, et al. (2005) Mapping with a Few Plants: Using Selective Mapping for Microsatellite Saturation of the Prunus Reference Map. Genetics 171: 1305–1309.

26. Cabrera A, Kozik A, Howad W, Arus P, Iezzoni AF, et al. (2009) Development and bin mapping of a Rosaceae Conserved Ortholog Set (COS) of markers. BMC Genomics 2009 10: 562.

27. Vendramin E, Dettori MT, Giovinazzi J, Micali S, Quarta R, et al. (2007) A set of EST–SSRs isolated from peach fruit transcriptome and their transportability across Prunus species. Mol Ecol Notes 7: 307–310.

Citations

CHAPTER 1

Zhao M, Xiang W, Tian D, Deng X, Huang Z, et al. (2013) Effects of Increased Nitrogen Deposition and Rotation Length on Long-Term Productivity of Cunninghamia lanceolata Plantation in Southern China. PLoS ONE 8(2): e55376. doi:10.1371/journal.pone.0055376

CHAPTER 2

Phalan B, Bertzky M, Butchart SHM, Donald PF, Scharlemann JPW, et al. (2013) Crop Expansion and Conservation Priorities in Tropical Countries. PLoS ONE 8(1): e51759. doi:10.1371/journal.pone.0051759

CHAPTER 3

Davis AS, Hill JD, Chase CA, Johanns AM, Liebman M (2012) Increasing Cropping System Diversity Balances Productivity, Profitability and Environmental Health. PLoS ONE 7(10): e47149. doi:10.1371/journal.pone.0047149

CHAPTER 4

Ray DK, Mueller ND, West PC, Foley JA (2013) Yield Trends Are Insufficient to Double Global Crop Production by 2050. PLoS ONE 8(6): e66428. doi:10.1371/journal.pone.0066428

CHAPTER 5

Aubertot J-N, Robin M-H (2013) Injury Profile SIMulator, a Qualitative Aggregative Modelling Framework to Predict Crop Injury Profile as a Function of Cropping Practices, and the Abiotic and Biotic Environment. I. Conceptual Bases. PLoS ONE 8(9): e73202. doi:10.1371/journal.pone.0073202

CHAPTER 6

Schmer MR, Vogel KP, Varvel GE, Follett RF, Mitchell RB, et al. (2014) Energy Potential and Greenhouse Gas Emissions from Bioenergy Cropping Systems on Marginally Productive Cropland. PLoS ONE 9(3): e89501. doi:10.1371/journal.pone.0089501.

CHAPTER 7

Casadebaig P, Quesnel G, Langlais M, Faivre R (2012) A Generic Model to Simulate Air-Borne Diseases as a Function of Crop Architecture. PLoS ONE 7(11): e49406. doi:10.1371/journal.pone.0049406.

CHAPTER 8

Lü H-Y, Liu X-F, Wei S-P, Zhang Y-M (2011) Epistatic Association Mapping in Homozygous Crop Cultivars. PLoS ONE 6(3): e17773. doi:10.1371/journal.pone.0017773

CHAPTER 9

Verde I, Bassil N, Scalabrin S, Gilmore B, Lawley CT, et al. (2012) Development and Evaluation of a 9K SNP Array for Peach by Internationally Coordinated SNP Detection and Validation in Breeding Germplasm. PLoS ONE 7(4): e35668. doi:10.1371/journal.pone.0035668.

INDEX

Q

R

S

T

U

V

W

Y